1/2/08

BUILDING A LUXURY CUSTOM HOME

What You Need To Know

BUILDING A LUXURY CUSTOM HOME

What You Need To Know

David A. Konkol
Master Custom Home Builder

BUILDING A LUXURY CUSTOM HOME

Published by House of Twelve Publishing
1000 N. Maitland Avenue, Maitland, FL 32751

General Editor: Todd Chobotar
Copy Editor: Jackie M. Johnson
Project Coordinator: Cindy Mallin
Author Photographer: Michael Lowry
Book Design: Chip David, Creative KX-3.Com

This book is a work of advice and opinion. Neither the author nor the publisher is responsible for actions based on the content of this book. It is not the purpose of this book to include all information about building a house. The book should be used as a general guide and not as a totality of information on the subject. In addition, materials, techniques and codes are continuously changing so please understand what is printed here may not be the most current information available.

This book contains numerous case histories and client stories. In order to preserve the privacy of the people involved, the author has disguised their names, appearances, and aspects of their personal stories so that they are not identifiable. Stories may also include composite characters.

ISBN-13: 978-0-9799736-0-4
ISBN-10: 0-9799736-0-0

Printed in the United States of America

9 8 7 6 5 4 3 2 1

First Edition

CONTENTS

Part II: During Construction

Part III: Helpful Checklists

ACKNOWLEDGMENTS

I'd like to thank my good friend Todd Chobotar, who asked me during one of our pre-dawn runs if I'd ever thought of writing a book. If it weren't for his encouragement, this book would never have been written. Todd is exceptionally talented in writing and editing, and I look forward to reading the first of his many books.

To Cindy Mallin, my personal assistant, whose love and passion is writing. Mine is not, and without her, my thoughts and structure would leave something to be desired.

To my friend Reed Collins, who listened to my stories on early morning runs, read the early draft, and helped me with many of the topics in this book.

To Kevin Howard, my Construction Manager of nine years, for his contributions and ideas. To my entire team at Dave Konkol Homes, whose hard work and commitment has allowed me the time, focus, and energy to concentrate on writing this book.

To Pam David, Karl VanDerAa, Nicole Bradley, Mike McDevitt, Tim and Sharon Peterson, Bob Brown, Dick and Carol Morrow, Jeff Wieland, Pat Snow, George Kiriakidi, Mike Dumont, and David Jacobs, who took the time to read the early draft of the book and offer their encouragement and helpful feedback. Your comments made it a better read.

To my clients, who have entrusted their hopes, dreams, aspirations, and often their life savings to me to build the home of their dreams. They've extended more grace than I deserve and I've learned many lessons along the way. Without their trust and willingness to come alongside me, I wouldn't have the experience and wisdom to share what is before you.

To my nine children—Lauren, Sam, Blake, Kate, Brantly, Alex, Anna, Andrew, and Jackson—who gave up a piece of their dad to allow me to write this book.

And most of all to you, Jo—my beautiful bride of 15 years, for your care, love, support, and encouragement when my days are long and the tasks before me are more than I feel I can handle. Your "You can do anything–I'm so proud of you" look is a source of tremendous energy. Thank you, Sweetie.

NOTE ON LANGUAGE USAGE

Please note: In order to achieve an easy flow of language, this book has been written using the singular pronoun "he" when referring to a builder. The author is fully aware that there are many competent builders who are women. The author has chosen to avoid the awkward use of "he/she" and has chosen instead to use the traditional masculine pronoun when referring to a builder. No offense is intended in this regard; the decision was made merely to achieve simplicity and flow of language.

BEFORE YOU BEGIN

Over the years I've had numerous conversations and have been asked countless questions about the things you need to know before building a custom home. To me, it's simple and straightforward because I've been in the residential home building business since I was ten years old. It's what I know best. But if you've never built a custom home or are considering building another one, then this book is for you. It's a quick read that gets straight to the point and provides the answers you need to make informed decisions. You can probably read the whole book in a couple of hours.

So why did I write a book on what you need to know before building your one-of-a-kind custom home? First, I wrote this book because I believe it can save you thousands (if not hundreds of thousands) of dollars. More importantly, it could save you a lot of headaches and heartaches. In truth, most of the insights in this book were discovered the hard way—learning from my own mistakes. And I'm still learning. My desire is to share practical information in an easy-to-read format to help you through the entire home building process (planning, designing and building) so you can create the luxury custom home of your dreams.

I first learned about hard work from my father. As a child, we lived on a dairy farm where long days of labor were a way of life. When I was ten, he sold the farm and started his own business building homes. I was always right behind my dad, following him around, handing him tools and helping in any way I could. I guess I got the entrepreneurial bug from my father because I started my own roofing company at age seventeen. In fact, that roofing company was profitable enough to put me through four years of college, where I earned a bachelor's degree in construction management. I've been building homes on my own ever since—and loving it!

If you ask someone who has recently built a custom home, you may hear a horror story about their experience. Perhaps four out of ten people will say their experience was "fair," and only one in ten will say their experience was "great." My aim is for you to be in the group that has a *great* experience building your new luxury custom home. I want you to imagine your dream home, to see it in your mind, and perhaps even sketch it out on a napkin over lunch with a friend because you're so excited about your unique vision. Then I want you to see it on paper, and some day, see your dream come true and live in the incredible home that you've created.

Of course, a house is more than just a structure. It's an extension of who you are. It's the place you raise your family, dream your dreams, and live them out to the fullest. Long after the concrete, wood, and glass make a *house*, the family makes a *home* and you come to cherish the place where laughter fills the air, sad moments come and go, and memories are made.

My deepest desire for the people whose homes I build is to someday—perhaps ten or twenty years later—be invited back so we can visit together and reflect on all the memories that have been created in this space that was once nothing more than someone's dream. It would be a treasure to feel the fullness in the air that holds all the laughter, tears, and memories of years gone by and share together what life has brought.

There are few things as personal as your home. Making the choice to build a luxury custom home can be one of life's most rewarding adventures. Without the right information it can also be filled with unforeseen problems and delays. I want you to have the best possible experience in your home building process. This book can be an essential and helpful resource. For the past fifteen years, my company has been building magnificent custom designed homes for the luxury market. The information in this guide, highlights from my years of experience, will help you make more informed choices, have less stress, and greater confidence throughout the entire process. Are you ready to get started?

Here's to building your dreams…

David A. Konkol

PART I

BEFORE YOU BUILD

Should I Build or Buy an Existing Home?

10 Questions to Help You Decide

"To build or to buy," that's the primary question to answer before building a new custom home. To help you decide, ask yourself these ten important questions. Be very honest. Answer each one carefully. Keep in mind that there are no right or wrong answers. You're simply trying to determine the best course of action at this point in your life.

Questions to consider:	(Yes or No)	
	☑	☐
1. Do I have a hard time making decisions?	☐	☐
2. Once I make decisions, do I struggle with wanting to change them?	☐	☐
3. Am I a perfectionist?	☐	☐
4. Is my schedule so busy it's difficult to find time to do the things I enjoy?	☐	☐
5. Does uncertainty and lack of control add stress to my life?	☐	☐
6. Am I regularly disappointed with interactions with other people?	☐	☐
7. Do I handle conflict by looking for the win/win solution?	☐	☐
8. Do I have some available time in my life for the next two to three years?	☐	☐
9. Am I realistic enough to recognize that things aren't always perfect?	☐	☐
10. Is our family life stable enough to handle the additional activity?	☐	☐

If you answered "no" to the first six questions, and "yes" to the last four, you're ready to build! If not, you may want to consider waiting on the building process. If your answers were different on more than three or four questions, I suggest you buy a house that is already built.

In my 20-plus years of building homes, I've seen many families stretched and stressed because the timing wasn't right or they weren't the right profile person to be building a custom home. They would've been better off buying an existing home than going through a process that wasn't suited for their life stage, temperament, or timing.

Let's look at an example of a couple who weren't ready for the custom home building process. In some ways, Josh and Melinda seemed to be ideal candidates to custom design and build a new home. However, as we got to know each other, I began to experience their difficulty in making decisions (see question #1). In addition, Josh was consumed by the demands of his business since he had just launched his own company three years before (see questions #4 and #8).

One day, Josh and Melinda arrived for an appointment in my office, late as usual. This time they were 40 minutes late, and by the time they arrived, they seemed more distracted than normal. I've learned that it can be difficult to have a productive meeting without a clear mind and focus, so I gently asked Josh if everything was okay, or if he needed a few minutes to switch gears. I know sometimes people just need time to transition.

Josh looked at me with eyes that seemed to say, "How did you know?" and proceeded to tell me about an intensely personal family situation that was consuming his attention (see question #10). At that moment, I realized that this lovely couple did not have the time at this point in their lives to spend custom designing and building a home. A quick look at the 10 questions above clearly indicated it wasn't an ideal time for Josh and Melinda to build a new home.

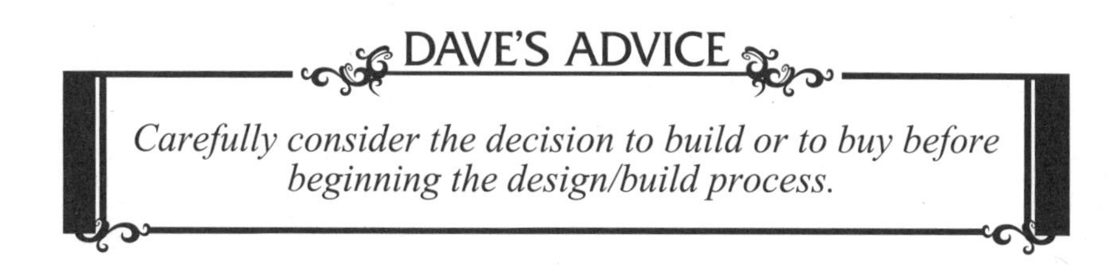

2 Which Comes First: The Builder or the Architect?

Just like the proverbial "chicken and egg" question, "Which comes first?" is a question that confuses some people, but must be answered before you start the custom home building process. While the answer may seem obvious, it's important to know in order to avoid problems from the beginning. The architect comes before the builder, right?

Wrong! Read on…

As an experienced home builder, I've met people who arrive at my office with a roll of plans tucked under their arm, a sparkle in their eye, and a skip in their step because they know they have something very special – the plans to their dream home. Like Mike and Janine.

During the last several months, Mike and Janine had spent countless hours dreaming about their new home and holding meetings with their architect. They went through revision after revision poring over the plans until late in the evenings. The couple worked tirelessly to make every room just right—put the baby's room here, move the daughter's room there, make that closet just a bit wider, add two feet to the kitchen—giving instruction upon instruction to their architect about each room.

Their dream home included the newest ideas from *This Old House,* the latest trends in low voltage lighting, and cutting-edge insulation that could lower energy bills by up to 90 percent. It had a cabana like the one they saw while vacationing in Acapulco, layers upon layers of moldings, extra tall ceilings, an additional bay in the garage, a steeper roof that was changed—not twice, but three times—because their friends told them it looked just too shallow.

And then the moment arrived.

Mike and Janine came to my office with wide eyes and great anticipation. Mike excitedly removed the rubber band from their rolled up plans, and spread them out to show me their dream home. This is a crucial moment for them; it's the result of months and months of their painstaking labor. They're ready to build!

For a few minutes I just listened. Janine was envisioning how much she'd enjoy the holidays in this beautiful home with their children, visiting friends and relatives. They could almost taste the good things their bright new home would bring—and then I did it. I asked the question.

"Do you have a budget in mind about what your new home is going to cost?"

Mike explained, "Well, the architect said we could probably build this home for this much per square foot, and if we leave out some of the crown molding, and if we do the painting ourselves, we will probably be able to find a builder who can do it for that amount." Both of them looked at me, expectant and hopeful.

And then, as kindly and as gently as I could, I told them the truth. From my professional experience, I knew that they were *75% over their architect's estimate*.

Silence.

Deafening silence.

Although I tried to say it in the gentlest way possible, it appeared I was the bearer of bad news. We continued for a few more minutes with polite pleasantries, then Mike and Janine rolled up their plans, walked out the door, and…sold their lot.

Their dreams had been shattered and they were crushed. After all their initial efforts, they couldn't gather the energy to start the process all over again. But it could have been different. If only this enthusiastic couple had known the importance of which comes first: the builder, not the architect.

If you're in this predicament and you're unwilling to sell your lot, turn the page and discover what happens next by reading Jim and Linda's story.

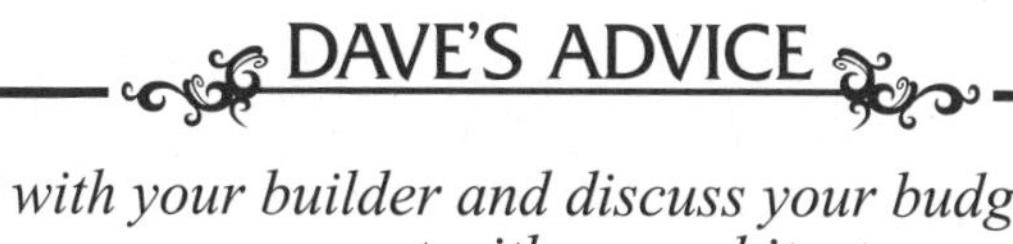

Meet with your builder and discuss your budget before you meet with an architect.

Forcing a Round Peg into a Square Hole

Finding a Competent Builder

Jim and Linda are the kind of people who refuse to give up. Sometimes persistence is a good thing, but there are times when pushing too hard is unwise. This couple, for instance, were unwilling to listen to sound, professional advice. They forced their opinions and ideas on a builder—and it was like forcing a round peg into a square hole. It just didn't work. Even when they realized they had received inaccurate advice from their architect about their home building costs, they wouldn't give up their dream.

So they began to shop in earnest for a builder who would build their home for the price they were told. Would Jim and Linda find a satisfactory and skilled home builder?

Maybe. But the builder they seemed to want—one who would be the answer to all their problems—would be either a builder who's desperate for work or one who doesn't know how to price a home.

Finding a competent builder can be challenging, but when you know what to look for, you'll get an accurate estimate and good advice. I've learned that many builders won't (or don't know how to) price a home while it's still in the concept stage. If most *builders* can't do this, I certainly don't expect architects to be able to provide an accurate estimate. After all, architects are trained and skilled at *designing* and creating what they are asked to create. Homeowners who don't have a good handle on pricing will tell the architect what they want and the architect will only do what he was retained to do. By no means am I blaming architects for not knowing about estimating accurate costs; it's not their area of expertise.

After dozens of exhausting interviews with many builders, Jim and Linda got their home building costs down to a price that was only 20 or 30 percent less than the initial estimate—still well over their budget. But they didn't want to give up their dream; they were willing to do anything to bring their dream to fruition.

At this point, Jim and Linda had some choices to make. Instead of cutting their losses and stopping the spending, they continued to pour more and more money into a project that wasn't suitable.

I've talked with other potential homeowners who received bad advice from a builder and learned, one or two years later, that their building project was a disaster. More than once, I've received a phone call in the middle of construction from someone asking me to get involved because they finally realized they had relied on poor advice and ended up involved in a lawsuit with their builder.

Please don't do this to yourself. Save yourself and your family the agony of lost time, lost dreams or lawsuits. Life is too short. It's not worth it.

Get good advice from a competent builder. Interview several first to determine the right one for you and your custom home project. A competent builder will explain the home building process and all the steps along the way. He can guide you through the entire process so you feel confident and secure in your purchase decisions. His firm will have an excellent reputation and be up-to-date on building codes, land, and procedures. Check out the company's references and previous projects completed.

Do I Have to Like My Builder?

4

So you've selected a competent builder. But you may ask, "Do I have to like the guy?" If he has a good reputation as a builder, does it really matter if I like him?

Yes, it matters. Don't sign a contract with a builder you don't like, trust, or respect. If you do, you could be headed for trouble.

Why? Because this is a long-term relationship and a long-term relationship with someone you don't like, trust, or respect can be challenging, frustrating, and more than disappointing. The planning stages of custom building a new home can take anywhere from months to years. Actual construction may range from six months to 24 months or longer, depending on the size and scope of your home. Add to that a one- or two-year limited warranty time period, as well as the fact that you may need additional information from your builder for many years to come regarding warranty information, vendor and subcontractor contacts, and other nuances.

In my 20+ years of building custom homes, I've lost some contracts to other builders and it usually boils down to *perceived* costs. A prospective homeowner may initially think our pricing is higher than our competitor, but most often that's because we didn't have the opportunity to thoroughly compare the two proposals.

We like to ask our homeowners why they chose Dave Konkol Homes to build their home. Almost always, the answer is trust. When they call and ask questions, they don't feel like we've postured a response that's politically correct. When challenges arise in your project, it's important to know that, when you call your builder, you'll get a straight and honest answer.

Do you respect your builder's values? I'm not suggesting that you have to socialize together, but I've seen people choose a builder they actually dislike. Maybe the husband likes the builder or his price, but his wife doesn't care for his style, approach, and manners. Ask yourself this question: Is there a reason you're uncomfortable with this person? If so, why in the world would you trust him to build your single most important investment?

If you or your spouse sense that a potential builder operates from a less-than-honest value system, why would you trust him to operate his business with honest values?

During the construction process there will be times when your builder will be making some judgment calls. Many of these will be unknown to you, and that's just part of the business. When it comes to your home, you'll want to know that your builder will be making choices, *as if it's his own home. As if his own family's safety depended on the choices he makes.* Not just what will pass code inspections. No shortcuts for a quick profit.

Are values important? You bet they are!

Don't sign a contract with someone you don't like, trust, or respect.

5 Pick Three Out of Four

Quality, Speed, Service, Price

My former neighbors, Jim and Susan, had six—count 'em, six—separate flooring companies at their home in a three week period providing estimates to sand and refinish their hardwood floors. About a month later, an unusual sound came from Jim and Susan's house, loud enough that everyone in the neighborhood heard it. I went outside to find out what the ruckus was about, and heard a diatribe of screaming and yelling that continued for several minutes. A hardwood flooring van was parked out front; I can only guess what happened. The couple expected a beautiful, high quality floor, but what they got was what they paid for. They weren't happy.

Jim and Susan had selected their hardwood flooring contractor based strictly on price, but they expected they would receive quality, speed, service, *and* the best price. Sure, they probably got the lowest price, but with it came a lot of heartache because they expected more and got a lot less.

It's no different than selecting a builder for your custom home. You need to determine what you value and decide what's most important: quality, speed, service, or price. Of course, you want all four components, but most often you will need to find a builder who can provide three out of four. That's reality.

Is it reasonable to expect that you'll get a builder who will give you the lowest price with great quality, great service, and a timely finish? Let's consider the merits of each:

1. QUALITY: A good company prides itself on providing a quality product, especially in the luxury custom home market. They encourage prospective homeowners to look closely at the work they've done for other homeowners and affirm they would be pleased with the excellent workmanship the company provides. Comfort with quality, opulence with outstanding craftsmanship.

2. SPEED: An on-time finish is important, but there may be times, especially in a busy market, when a builder misses some deadlines. If that, happens, you want your builder to proactively communicate with you and, if possible, find a way to make up the time and get it done quickly.

3. SERVICE: A builder with exceptional customer care will provide good communication and attend to the homeowner's needs before, during, and after their home is completed.

4. PRICE: The best companies aren't usually the cheapest, but consider this: they're probably not the most expensive either. Great companies deliver good value. An honest builder charges at or below market value for the level of service and quality product they provide. Very seldom is the cheapest price the best choice for a home builder.

Don't make the mistake of thinking you can have all four qualities in one builder. A Lexus or Mercedes is priced differently than a Pontiac. If you pay Pontiac pricing (like Jim and Susan) and still expect a Mercedes or Lexus level of performance, you are setting yourself up for disappointment, conflict, and sometimes even a lawsuit.

Know what's important to you and adjust your expectations. If cost is your most important value, then choose the contract with the lowest price. However, if you value quality, be sure you look for excellent workmanship. You can get the results you want in a luxury custom home; just be sure to select your builder based on what you truly value.

Realistically expect to get three out of four components from your home builder, so decide what's important to you.

Where Should I Spend My Money?

If You're Going to Err, Do It Here

Building a home is probably one of the biggest investments you'll ever make, so you'll want to know where to put your money to get the most value for your investment. Here are seven areas to consider:

LOCATION: You've heard the adage *location*, *location*, *location*. This is where to spend your money. You could build the Taj Mahal Palace, but if you build it in the wrong location or on the wrong lot, you could be in real trouble. I always tell people to buy the most expensive lot they can afford. Historically, waterfront property experiences stronger growth in value than non-waterfront property. If you like water and can afford it, build on the waterfront. Your long-term investment is more likely to be sound.

DESIGN SERVICES: You can add enormous value to your home by investing in the services of a competent architect and an interior design team. Not only will you enjoy the splendor of a fabulous home, you will find a greater return on the money you spend for these services at the time of resale.

KITCHEN: In my entire career of building homes, I've never heard anyone say that a kitchen was too large, there was too much counter space, or too many cabinets. Spend money on the kitchen. We'll talk more about this topic later.

FAMILY ROOM: Oversize, don't undersize your family room. Families tend to gather and spend most of their time in the family room. If I were to oversize any room, I'd make this room a little larger (rather than smaller) than you think you need.

MASTER BATH: This room is the owner's retreat, a place to relax and unwind. Upgrade your master bath's size and finishes. When you sell your home, this will be an important feature and provide a good investment return.

ROOM SIZE: Make sure your rooms are large enough to meet your needs. It's very expensive to come back after your home is finished and add twelve or eighteen inches to a room because you've just realized that it's too small. If you're on a limited budget, it's better to hold off on some of the finishes than cut down the size of your rooms. Later, you can add finishes and the cost may only be slightly higher than if you installed them during the initial construction process.

CLOSETS: Never underestimate the value of roomy walk-in closets, linen closets, and laundry rooms.

DAVE'S ADVICE

Invest your money in the seven places that matter most and you will experience great value for years to come.

7 What's So Important About the Kitchen Anyway?

When a couple buys or builds a home, they always seem to pay a lot of attention to the amenities in the kitchen. Within most couples, there's usually one person who loves to cook; most often it's the woman. But with everyone, male or female, I've found that the kitchen is one of the most important rooms in the house. The kitchen is often the hub of the home, the center of activity. Someone spends time preparing food, creating something fun, or trying a new recipe in it. People tend to gather where there's food.

So don't miss this: kitchens are important!

Men sometimes underestimate the importance of this room. We love our garages, grills, decks, and patios. But the kitchen? We can completely miss its importance. I always counsel people who are designing a home to remember the resale value, and design accordingly. A well-designed and functional kitchen not only makes the cook happy, it also ensures a greater resale value if you sell the home.

A person who loves to cook or create culinary masterpieces for family and friends absolutely needs a spacious and well-appointed kitchen. On the other hand, if a cook wants to spend as little time as possible in the kitchen, then your design and layout can be simpler. My friend Kathy doesn't spend much time in her kitchen, but she likes the spacious layout so she and her husband can be there together. Kathy says the chopping and stirring are more enjoyable with her husband. Additionally, he likes the organization of a well-planned room.

When selecting kitchen cabinets, look for quality. This is not the place to be overly consumed with trying to save money. Well-made cabinets will provide lasting pleasure and functionality for decades. Drawers (as opposed to cabinets with doors) are more useful and efficient even compared with cabinets with pullout drawers. In addition, choose hard surface, durable, and high quality countertops. There are a variety of colors and selections available. Granite is an ideal counter surface for the most important room in the house.

What about appliances? Don't skimp here. I'm not advocating that you have to purchase the absolute top of the line, but good quality appliances help sell homes. Today it's becoming very popular to have two dishwashers. Often the cost of an additional dishwasher is only slightly more than the cost of the additional cabinetry it replaces.

In order to make the most important room in your house spectacular, consider adding some of these other features: warming drawers, double ovens, convection microwaves, pot fillers, espresso and coffee bars, hidden pantries, elevated dishwashers for easy access, vegetable sinks in the island (in addition to the main sink), instant hot water dispensers, and purified water faucets.

When you design your home, be sure the kitchen is given special consideration and that its relationship to other rooms, function, and features are the best you can provide. It will make a difference for years to come.

DAVE'S ADVICE

Don't treat your kitchen like any other room in your home. This is one room that deserves special treatment.

Needs Versus Wants - The Design Outline™

8

One of the best lessons I learned early in my career was to require homeowners to thoughtfully consider the difference between needs and wants. I remember meeting with Tim and Sherri, a couple who debated for 45 minutes on whether or not they wanted a formal living room. As I sat there thinking about all the other things I needed to be doing, I figured there must be a better way. So I developed and trademarked the **Design Outline**™.

The **Design Outline** is an excellent tool to help you define your needs and wants in the home building process. This exercise takes less than 30 minutes to complete, but it can save you countless hours and thousands of dollars. For an example of a completed **Design Outline** form, see Appendix II. Here's how it works:

1. You and your spouse, independently of each other, each take a blank sheet of paper and write down all of your dreams, wants, and needs for your new home, in no particular order.

2. Then rank your items in order of importance, starting with #1, #2, #3, etc. It's not as important to agonize over whether item two goes before item three, or three before two; sometimes wants or needs can be equally important to you. What's important here is that #3 and #28 are not reversed. Know what you want.

3. Once you and your spouse have independently ranked your items, the two of you meet together to share and compare your lists. Then create one combined list ranking your needs and wants in order. This will become your master list.

The combined needs/wants list will save time, energy, and money when you meet with your builder to help you determine the cost of your new dream home. At some point, your desired budget will need to line up with your desired wants.

Your builder can review this combined list and your budget, and let you know what items your budget can afford. If you have items that are not included in the budget, your builder can estimate a value so you can make an informed decision on whether or not you want to increase your original budget.

The Design Outline has been a tremendously effective tool that we've used to help people determine costs long before they spend money on design or construction of their new custom home. You can use the Design Outline with any builder, anywhere in the country. When you begin the process by implementing the outline I've created, you'll be way ahead of the game.

Take 30 minutes and complete the Design Outline.
It's fun and it will save you countless hours and thousands of dollars.

Should I Use a Fixed-Price or Cost-Plus Contract?

Part 1 – Fixed Price

A fixed-price contract is one in which the plans, specifications, and all of the materials and finishes are fully determined (fixed) before you start construction on your new home. A cost-plus contract involves taking all the costs of the home and adding a percentage of costs or a flat fee for the builder's overhead and management fee.

In my experience, either a fixed-price or a cost-plus contract can be used successfully in building a custom home. It depends on what you're most comfortable with. In my business, I complete approximately 50 percent of our homes on a fixed-price contract and 50 percent on a cost-plus contract. Your builder's costs are the same whether he uses a fixed-price or cost-plus contract. However, the price you'll pay will differ because of the risk associated with each type of contract.

Let's start by looking at a fixed-price contract.

The advantage of a fixed-price contract is that the price you pay for your home will be predetermined (fixed) whether or not the price of material and labor goes up or down. Your builder assumes full responsibility for all risks associated with the cost of your new home. The downside is that you'll pay more for your builder taking on this risk.

On a fixed-price contract, the builder assumes responsibility for all risks associated with the fluctuation in costs. Of course, costs are always changing. From the time you sign your contract to the completion of the work, the actual costs will change. Sometimes up, sometimes down. If labor and material go down from the time you sign the contract until the job is complete, the builder benefits. If the cost of labor and material go up during the construction period, the builder absorbs the loss.

Another factor in pricing a home with a fixed-price contract involves what I call the fright factor: will my homeowners change their demeanor once construction begins? No one has ever built a perfect home without some sort of scratch or blemish on it. I can walk into any home in America and find something that's wrong. I can take a magnifying glass and find scratches on any window on any newly built home in America.

Therefore, builders need to charge clients for "fright." If the builder gets to the final walk-through with a homeowner who takes out a magnifying glass and searches every square inch of every window pane in the house, finds a scratch, and wants every scratched window replaced—at my expense—I'd better have "fright" included in my initial calculations. With a fixed-price contract, the cost of repairing or replacing any and every item that has even a tiny imperfection is all the builder's responsibility.

Over the years I've found that people with a fixed-price contract are more inclined to expect imperfect minor items to be replaced because it doesn't cost them extra. To be clear, I'm not talking about shoddy workmanship. I'm talking about the gray areas of requests that are unreasonable, based on industry standards.

In general, people who do well with a fixed-price contract are people who are not willing to risk price fluctuations. They are more comfortable knowing their exact cost and are willing to pay a small premium for this comfort. In the next chapter, we will cover the option of a cost-plus contract.

Choose a fixed-price contract if you are willing to pay a small premium for locking in the total cost of your contract.

Should I Use a Fixed-Price or Cost-Plus Contract?

Part 2 – Cost-Plus

A cost-plus contract differs from a fixed price contract in that it takes the actual cost of building the home and adds a percentage for the builder's overhead and management fee. This fee can be either a lump sum or a percentage of total costs. Of course, neither you nor your builder will know the exact bottom line for the building costs until the final accounting is completed shortly after closing. A good builder will give you an accurate cost estimate, but it's exactly that—an estimate—until the final accounting. You'll pay the actual costs for all labor and material, plus the builder's fee.

A cost-plus contract can be advantageous when building larger custom homes with finish levels and other things changing during the process. With the cost-plus basis, the homeowners know their actual costs on an ongoing basis. They can then determine where they will appropriate their funds early in the construction process, and receive a full accounting disclosure of all costs. If they elect to pull out a magnifying glass to search for scratches in all the panes of glass, they can choose to have those panes of glass replaced at their cost.

Usually, doing business on a cost-plus basis keeps the magnifying glass in the drawer. It doesn't mean the builder builds with less care or quality; it just puts the homeowner and the builder on the same team. A cost-plus contract provides the synergy of identifying problems and determining win/win solutions that are in the best interests of the homeowner.

If you trust that your builder is competent and is working on your behalf, and if you are comfortable not knowing your exact total costs until the end of the project, then a cost-plus contract may be best for you. Your final cost will depend on the choices you make. At our company, we charge a smaller management fee on a cost-plus basis because we incur less risk.

Both fixed-price and cost-plus contracts are successfully used in building new custom homes. You can decide which one works best for you

DAVE'S ADVICE

Choose a cost-plus contract if you are comfortable knowing the end cost will be determined by the choices you make and you know you have a trustworthy builder.

Don't Even Begin Designing Your Home
Until You've Done This!

11

One of my favorite sayings comes from the Bible, a book filled with wise truths. In the book of Luke it says, *"Suppose one of you wants to build a tower. Will he not first sit down and estimate the cost to see if he has enough money to complete it?" (Luke 14:28, NIV)* I suppose it's because I'm a builder, but I can relate to this thought. It's a rhetorical question, of course. Who *wouldn't* first estimate the cost?

In reality, it sometimes amazes me to meet with people who are ready to build a new custom home, but have no idea what they want or how much they can spend on it. Stop right here! It's crucial, before beginning the design-build process, to determine what you can afford.

We refer people who need financing assistance to an experienced professional mortgage lender who can help determine what they can comfortably afford. This process takes into account your income, expenses, credit, assets, interest rate, taxes, insurance, maintenance, and utilities.

Sadly, people who begin the design process without first counting the cost often design to their dreams, only to find out later they have far exceeded what they can afford. They end up mad, sad, or extremely frustrated. On the contrary, we want our homeowners to first realistically determine what they can afford; then we work hard to design a luxury custom home, a complete package, that's 5 to 10 percent below their target number. We know from experience that changes may occur during the building process. For instance, a homeowner may upgrade the finishes as the construction process unfolds or other variables may arise that would add to the cost of their home. If we start with a number that is below their budget, we can end up at the desired budget.

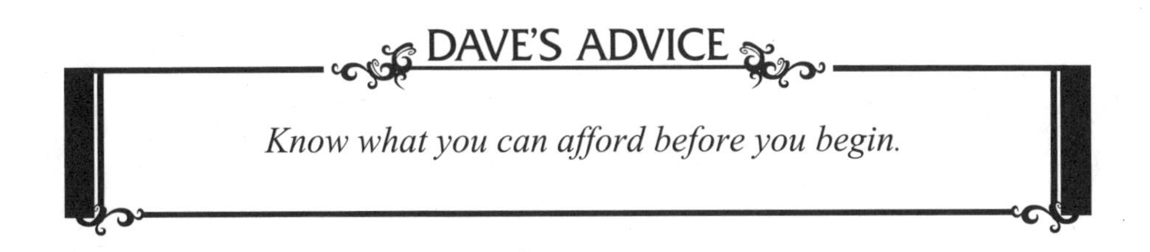
DAVE'S ADVICE

Know what you can afford before you begin.

Buy A Plan Or Custom Design?

Read This Before You Decide

Should you buy a ready-made plan for constructing your new home? It depends.

If you have a large lot that doesn't have restrictive setbacks, then finding and buying a ready-made plan may work well for you. It's important to know that most plans you find in a plan book are designed for specific lot sizes and regular shaped lots. If your lot is long and narrow, and the plan is designed for a lot that's narrow and long (just the opposite), you could have a problem.

Sometimes people will use what I call the "cut and paste" method. A couple will take a plan from a plan book and "cut and paste" it to make it fit their lot. Be careful with this method. Judy was a woman whose cut-and-paste plan didn't work out the way she'd planned; it required you to walk through one bedroom to get to another bedroom.

If you can find a plan that fits on your lot with little or no changes, it may be a good option for you. However, I've met with very few people who say they are completely happy with a design they found in a plan book without making minor or major modifications.

If you are thinking of buying a ready-made plan and just shrinking it, it almost certainly will *not* work.

Our company has successfully built homes from ready-made plans. We purchase the rights for the plan, make any needed modifications, and have a successful build. Most good custom home builders could do the same for you.

For people who don't choose a standard plan, we offer a complete design/build package. The process starts with the shape of the lot; it's important to take into account its positive and negative features. Then, we take our homeowner's completed Design Outline™ (see Appendix II) of their prioritized needs and wants and begin the design process. Our initial meeting with the owners is an important step to review the design and to ensure we have a complete and thorough understanding of their needs, wants, and budget. Once we've clearly identified the objectives, we schedule a series of three-hour design meetings with our in-house licensed architect to create an elegant, luxury custom home—the home of their dreams.

We've found that it doesn't necessarily cost any more (and often we've discovered cost savings) by first focusing on and designing what our homeowners most *need* while also taking into account what they *want*. In doing so, we've found great success because the process allows us to optimize the conditions of the specific and unique lot instead of having to work from an existing design by enlarging or shrinking areas in the home. With clear objectives and a good plan, you can focus your energy and resources on your unique needs for your new home.

DAVE'S ADVICE

If you can find a ready-made plan that requires few or no changes and you just love it, then you may want to consider purchasing a buyer-ready plan. If not, don't make the mistake of cutting and pasting.

13 Surprises That Could Inflate Your Custom Home Cost

Early in the design process, your builder should work with you to create an estimated line-item budget for your home. Each item (landscaping, appliances, flooring, etc.) should be listed separately with a corresponding dollar amount (allowance). The line-item budget should make sense to you, and the numbers should be in line with the general caliber of work you've seen in other homes this builder has completed.

Interestingly, an allowance (or budget) can be your friend or it can be your worst enemy. In the past, our company has lost jobs to the competition because I think it's important to establish a realistic allowance that truly reflects the scope of the entire job. Conversely, many builders will establish low allowances that are insufficient to build the home with the quality expected. Be wary of a budget that seems too good to be true. You could be talking to an unscrupulous builder who says whatever he has to say to get your business.

Allowances are established by builders because selections are not always made prior to the commencement of building a new home. If a builder provides an estimate for a new home and purposely or unknowingly lists allowances that are not sufficient for the quality expected, the initial bid can be thousands, if not hundreds of thousands, less than our bid proposal.

The last thing my homeowners or I need is to be in the middle of construction with insufficient allowances to be able to complete the home. We put a lot of effort into listening to what a homeowner values and then designing a complete home package to meet those needs. We try to establish budgets (allowances) that allow our homeowners to select items that are commensurate with the level of quality they want for their home.

One of the biggest challenges for a homeowner and a builder is to identify what the sufficient allowance should be and how all those numbers and allowances play into an overall budget. I believe that if you give someone enough time and money they can accomplish anything. A major key to a successful project is to take the desired overall budget, consider the needs and wants of the homeowner, and create and execute a home that the homeowner will be happy with for years to come.

After listening to the needs of our homeowners, I establish allowances that I think are appropriate for that home. After that, we encourage homeowners to do some shopping at our selected vendors before they sign a contract. This will allow them to get a sense as to whether their allowances are sufficient or not. If the allowances are not sufficient, we either have to raise the overall budget or reduce allowances or features in other parts of the home. It's important to do this *before* construction begins. If determining allowances happens *after* construction begins, it leaves room for misunderstanding and frustration.

DAVE'S ADVICE

Know what your allowances will buy for your new home before you sign a contract.

14

Design Your New Home With Resale in Mind

Who thinks about resale value when building a home? While it may seem odd for a person who's building a new home to think about selling it, it's an important point to address what you want (and what the market wants) even in the initial planning stages. If you're not careful, you may design your dream home and find that no one else will buy it later!

Most people we build for are financially sound. For the most part, they are able to build multi-million dollar homes because they've been making good financial decisions for many years. As you go through your design process, be certain you get wise counsel from your builder, architect, and possibly a real estate salesperson, to be sure you're not building a home that only your family will like. Balance your wants and needs for the home of your dreams with potential market appeal for future resale.

My homeowners will often hear me say, "I know you're not building this as a spec* home, but this particular design feature is not what the broad market is asking for." As an experienced professional, I feel a responsibility and duty to share this crucial information. Once you have the facts, it's your responsibility to make your own design decisions. If you proceed to design and build a home that only appeals to a very narrow market, then at least you are aware of that (and the consequences) in the early stages of the design process.

The fact is I've seen many homes linger on the market for a long time because a homeowner made design decisions without taking into account the long-term resale effects.

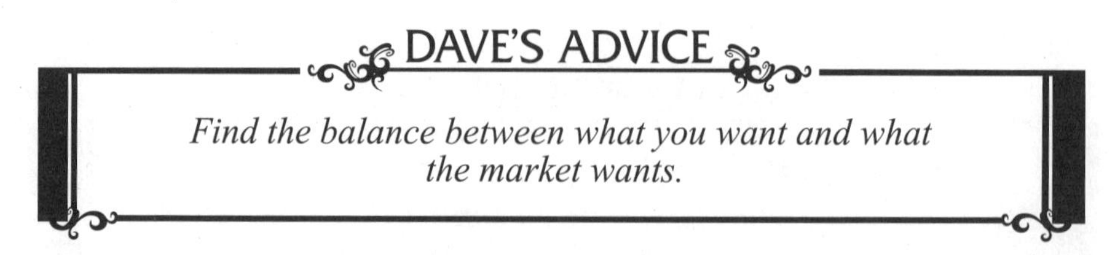

*A spec home is a home without an end buyer before construction begins, generally built with an intended profit.

15 You're Going to Live Here HOW Long?

During my first conversation with someone new, I often ask, "How long do you intend to live in this home?" And I sometimes hear, "This is the last home I will ever build. They will have to bury me in the back yard."

Recent surveys show that most people live in their home an average of five years and you're probably not much different. You may be thinking, *But Dave, this really is going to be my last home.* And if it is, that's great. But you may want to consider the reality of averages.

Susan took more than a year to design her new home for her family only to discover she was way over budget. She also realized that, by the time she completed her home construction, all but one of her five children would be away at college, and she had designed bedrooms for each of them! Reality finally dawned on her that, within a few short years, she and her husband would be empty-nesters. Designing her home for a family of seven thinking this would be her home for the rest of her life didn't fit the reality of Susan's imminent transition to a home for two.

When designing your custom home, first design for your immediate needs. Second, take into account what market conditions will allow for your particular home design. Third, give strong consideration to how long you may live in your home. Susan may shift gears and design a home that's perfect for her family's needs for the next five to seven years with a plan to downsize at that time. With this more realistic outlook, she may give additional consideration to the resale value of her choices.

Don't make the mistake of over-designing a home that may become obsolete for your family within a short period of time.

Be realistic about your short-term and long-term needs and how long you will live in your home, and design accordingly.

16 Should I Hire an Attorney?

If it makes you feel better, go ahead—hire an attorney. But if you do, hire someone who knows the real estate and construction business and will complement the process.

Monica first came to me because I was building a new lakefront home just a few doors down from her. Most Saturday mornings I'd see Monica walking the new home, inspecting the quality of workmanship and materials. Monica not only knew my homeowner, she knew a number of my previous homeowners.

As Monica and I got to know each other, we began planning to build her a custom home. We spent many hours and numerous meetings designing the ideal plan for her lakefront lot. Finally, two-and-a-half years later and after three re-dos of her design, we were ready to build.

Three days before the closing date on her loan (signaling the beginning of construction) Monica called me to say her attorney had reviewed my contract and advised that it needed to be completely rewritten. I told Monica I'd been using this contract for over 15 years and had never experienced a request like this. I suggested she ask her attorney for specific comments and told her I was confident we could work through the issues. Monica agreed and said she would get back to me.

The day before the scheduled closing I received an email from Monica stating that (based on her attorney's advice) she would not be building with me. Obviously, I was surprised and shocked.

A few days later, I met with Monica and she affirmed that, following her attorney's advice, she would not be building with me. I told her I was disappointed because we'd been working well together for almost three years. I had counseled her through obstacles, given her professional advice, and taken her through three design changes— all successfully. It was very disconcerting now, after all that, to learn that her attorney was counseling her *not* to proceed without even a willingness to discuss specific points.

Unfortunately, Monica received some bad advice from her attorney. By insisting on an entire rewrite of the contract a few days before closing, he didn't serve his client well.

Monica eventually built a home (with another builder) and it was close to my neighborhood, so every few months I would drive past the project. A good builder can tell without ever speaking to anyone whether a project is going well or not. By seeing the progress over the months, I observed that the project took six or seven months longer than anticipated. It didn't seem to be a good experience for the builder or for Monica.

I'm not bashing attorneys here; some of my best friends are attorneys. In fact, I have a great attorney who charges me nearly $400 an hour, but he provides essential advice that complements who I am and what I do. He helps me design win/win agreements, not win/lose or lose/lose situations.

If you hire an attorney, be sure to hire someone with construction and real estate experience, not a general practice attorney who counsels on personal business, family trusts, estates, etc.

Hire an attorney to review your contract if it makes you feel better. Just be certain to hire one who helps you and not hinders you.

17

How Much Do You Charge Per Square Foot?

Building a custom home is a bit like buying a new car, right? Not exactly.

If you asked a car dealer how much he charges *per pound*, you'd get some very strange looks. Of course, there is some correlation between the cost of the car and its weight, but not significant enough to prompt that question. We all know car dealers don't sell by the pound.

In the same way, I feel perplexed when someone asks me how much I charge *per square foot* to build a home. It's not the right question.

There are three factors that contribute to the cost of a home, regardless of where it's built: complexity, level of finish, and size and components.

1. COMPLEXITY: A home with more features and greater complexity requires more labor, and therefore costs more to build. For example, a rectangular house with four basic corners is less expensive to build than a three-story home with 40 corners, angled walls, and steep roofs because the latter is more complex and takes more time to complete.

2. LEVEL OF FINISH: Obviously, vinyl flooring is much less expensive than wood or stone. Formica countertops are less expensive than granite. Twelve-inch baseboards cost more than six inch baseboards and a lot of molding is more expensive than no molding at all. The level of finish you choose for your home will have a significant impact on the home's final cost.

3. SIZE AND COMPONENTS: Size matters in home building costs. A 6,000 square foot home will cost more than a 2,000 square foot home. A 2,000 square foot home would probably include a two-car garage, while a 6,000 square foot home normally has three or four bays. So, not only does the larger home cost more due to the size of the air-conditioned space, but it also takes into account things like garages, number and size of porches, whether the home has a pool, boat dock, circular drive, and other costly components.

It's a good idea to ask a builder what price **range** per square foot he builds at, in order to know if you're talking with the right builder.

DAVE'S ADVICE

Don't go to a Chevy dealership if you want to buy a Lexus.

18 Welcome to Art Class!

Create a Dream Home Notebook

Now you get to have some fun! While you continue to dream about the possibilities for your new luxury custom home, let's create a "Dream Home Notebook." Gather some magazines and tear out pictures of features you really like or want in your home, and just like you did in art class, start collecting them in a notebook. Write on each picture specifically what it is you like about it. Why did you tear it out? People often bring me pictures, but months later they can't recall why they tore it out in the first place. Write it down.

Creating your Dream Home Notebook can be an inspiring and enjoyable activity, and it will be very helpful to your builder and design team. As you accumulate more and more pictures, begin to categorize them. Here are some suggestions for your categories:

1. Exterior features.
2. Kitchen features.
3. Master bathroom and bedroom.
4. Common living spaces.
5. Specialty items such as fireplaces, mantles, trim details, paint colors, etc.
6. Colors, textures, styles.
7. Any floor plan that interests you. Write down what interests you about a particular plan. It may be the relationship of rooms, the uniqueness of design, or even a small feature like a hidden pantry or a workstation for mom.
8. Things you specifically DON'T like.
9. Your written notes.

Sometimes it can be difficult to express what you like and don't like in new home features. So the Dream Home Notebook is a helpful resource in the planning process. The saying, *"A picture is worth a thousand words"* also applies to the development of your custom home design. As an experienced builder, I can look at the pictures and listen to the conversation (and often read between the lines) to help you better articulate what you are thinking but may be unable to put into words.

It's also helpful to write down your expectations. You may want to include stories (good and bad) of what your friends have gone through in their home building experience. Also write down the elements in your home that are important to you. When you finally meet with your builder, you will have a well-organized, thoughtful notebook to share, which will help tremendously in your design and building process.

Cut out pictures for your dream home and organize them in a single place along with your notes on what you really want.

Close Enough to Perfect?

Identifying Expectations

A home builder's goal is to create a well-built, attractive home that meets the needs of the homeowners. But how close to perfect does the finished product need to be? What about imperfections or flaws? My friend Brian doesn't care about details, while George is extremely particular and wants every single blemish erased—*every single one.*

What about you? What are your expectations for what you will and will not accept when your home is finally completed? This is definitely something you and your builder need to discuss.

Here's what I often do: I make an appointment to meet the new prospective homeowner at one of our recently completed homes. I have a very defined plan to walk through this new home together and thoroughly inspect all aspects of the home. We take about one hour of uninterrupted, private time without the owner present. The reason I don't want the owner present is because I've found that if the owner is there, the prospective homeowner is reluctant to look very closely at the fine details.

I meet the prospective homeowners at the newly-built home and start the inspection in the foyer. As I point out the features, I ask a very specific question. "If we were standing here a year from today and doing the final walk-through on *your* new home, would this meet your expectations?" Usually, my wide-eyed prospective homeowners nod affirmatively. Before I move on to the next room, I take a moment and encourage them to look closely at the finish. I tell them that a paint job can never be perfect; I may find a blemish or two on the walls or in the trim work. Then I ask again, "Would a paint job with these imperfections meet your expectations?" Usually, they say yes. Then I ask them to rub their hands over the trim work to feel for smoothness and any imperfections. I'm actually looking for some imperfections in the paint job because I want to clearly identify their expectations.

After we discuss the paint finish, we move on to trim work. From trim work, we move to drywall. From drywall, we move to flooring, and on and on it goes throughout the house. Then we walk into the living room where the same process takes place. I ask the prospective homeowner the same question, "If this was your living room, aside from the paint colors and selections of materials (which will probably be different in your home), would this living room and the quality of the workmanship meet your expectations?"

This process is very important. The last thing you or your builder want is to find yourself a year or 18 months into the building of your new custom home, only to discover that each of you had different expectations. I make it a point during the interview process to be sure we have clear expectations of what is or is not acceptable for the quality of the end product–your luxury custom home. By doing so, both parties can avoid unmet expectations, frustration, anger, or even a lawsuit.

Schedule a one-hour private showing of a home your builder has recently completed. Tell your builder whether the quality meets your expectations or not.

20 The Terrible Truth About Building Beyond Your Means

Sometimes I work with homeowners who want to build a home that doesn't seem to be affordable for them; it's well beyond their means. I struggle with that double-edged type of situation because if I build the home the way they want, I know they'll regret it. On the other hand, if I don't, they won't be happy with me. In fact, I've sometimes lost building opportunities simply because I believe it's my obligation to be straight with people when it appears they're pushing the envelope of costs.

Of course, I know I'm not my client's keeper regarding how they spend their money. But as a professional who values integrity, I believe I have a responsibility to share the cold, hard facts of the large investment they will be making in building a home. Some people begin the design process with a realistic budget that's within their means, but as the process moves along, it can begin to get out of hand.

If you've purchased a new car recently, you know what I mean. Say, for example, you want to get a nice car for $37,000. Once on the lot, you see the base price on a model you like is–$45,000. You really like the upgraded 10-disc CD changer (only another $1,000), and it's just $19.80 a month more on your monthly payment. Of course, then you see other cool things like the GPS Navigation System, the backup camera, and the Premier Audio System—so you add another $6,000. You also decide to upgrade from the standard leather package to the heated and cooled, comfort leather seats. That adds $2,160, but it's so incredible! Then you discover the Satellite Radio System, and add $486. Finally, you decide to add a sunroof, custom paint, and upgraded Pirelli tires…

You get the picture.

In a matter of minutes, your $45,000 car became a $62,000 car. And remember, you started out looking for a car that would cost $37,000.

That's what it's like for some homeowners who design and build a home. Construction hasn't even started and already they've added options and selections to their home beyond the original plan. I've seen people stretch and stretch financially to build their new home, and by the time construction begins, they're under so much stress that when we have an opportunity to add a nice feature to their home (something as simple as additional crown molding), they are completely stressed out because of a few hundred dollar decision. That's often because they didn't discipline themselves to stick to a reasonable budget.

Please don't build a dream home you will have to sell before you even move in because you can't afford it.

Be sure you have a builder who will help guide you through this process with honesty and professionalism. If I notice that costs are beginning to exceed the homeowner's budget, I tell them that it's my responsibility to communicate to them about costs that may exceed their budget. The final choice is up to the homeowner. Ultimately, I want them to be happy *and* financially healthy with their finished dream home.

Build within your means so your dream home doesn't become a financial nightmare.

21 The Hidden Pot of Gold in Your Pocket

When I work with people on a home building project, sometimes couples will come to me with a maximum budget for building their new home. They plan a budget for the lot, the home, and architecture—a complete package. When I review it with them, I spend a considerable amount of time listening, asking questions, and working to meet their particular budget. I want to make sure we maximize their investment.

Many times, however, I'd work fervently to create a design, size, and features to meet the homeowner's needs and wants within the budget they presented, only to find out later that they had a hidden "pot of gold" in their pocket; they had more to spend than they let on originally.

That approach has always seemed counterproductive to me. After spending countless hours working to meet their budget, and THEN learning the couple actually has an additional amount to spend, we essentially have to go back to the drawing board to start the design process all over again. Not only is it frustrating, it also sets the construction schedule back. So if they thought they'd be in their new home by Thanksgiving, it would now be closer to the fourth of July. After this hidden amount situation happened to me several times, I started telling my homeowners this true story:

When I first got out of college, I was ready to purchase my first vehicle. I went to the car lot and told the car salesman I wanted to spend no more than $10,000 for a vehicle. He said, "Great. Come out to the lot and I'll show you what I have."

I followed him out, and when he showed me the first car he said, "Now here's one for $11,500." I looked at him in complete disbelief. I thought, *Did you not hear me? My budget is $10,000.* I was thinking he'd start me me somewhere in the $8,000 or $8,500 range, so that by the time I paid for tax, title, car handling fee (to this day, I haven't figured out what that fee is), and other options, I'd end up at my budget of $10,000.

Admittedly, I was young and naïve and didn't know about slick salesmen. I did end up buying a car from him because I didn't know any better, but it gave me a bad taste in my mouth. I told myself then that if I ever had the opportunity to sell someone a product or a service, I'd never sell that way.

I tell homeowners this story because I finally realized–they may think I'm like a slick car salesman. So, I tell them I treat their money like it's my own. I will honor and take seriously the budget they give me. That's why I need the budget to be accurate, with no mysterious "pot of gold" showing up later.

If you trust your builder, give him your real budget at the beginning of the process. If you don't trust your builder, then you shouldn't be working with him anyway.

DAVE'S ADVICE

Hire a respected builder, then trust him with your real budget.

Choices, Choices, Choices!
Choose Before You Lose

22

I strongly recommend making your selections before construction begins on your new custom home—color, fabric, plumbing, hardware, paint, and more. Yes, – all selections!

Before I learned how important this was, I noticed that homeowners often seemed unable to make selection decisions in a timely manner. Indecision messes up the deadlines, stalls the project, and can greatly frustrate both builder and homeowner.

Randy is a good example. When I built his custom home, I gave Randy deadlines for his selections and every single deadline was missed. I called him and told him it was time to make his paint selection, and he asked me, "Well, when do you need it?" I said I needed it two weeks ago and he said, "But when do you absolutely, really have to have it?" I told him "Friday," and he promised me he would have the paint selected by Friday.

Well, Friday came and went with no paint selection. I called Randy on Monday, and he told me he had an emergency, which precluded him from making his paint selection, but he would have it to me by Wednesday.

At about 4:00 p.m. on Wednesday, Randy showed up and began to put samples on the wall only to find out he needed more samples to compare to his original samples. The next day more paint samples went up. Another Friday passed and Randy said he was having a difficult time making a decision.

In the meantime, we had already completed the drywall and work on the home was at a standstill. The process had lost momentum. I was frustrated, Randy was frustrated, Randy's wife was frustrated, and the painter was frustrated. Even the cabinet man was frustrated because now his schedule was delayed.

Everyone was frustrated and all the work had stopped.

The painter wasn't sure he wanted to sign up for my next job, and my reputation as a builder started to get a little shaky.

Finally, I found a way to prevent future messes like this. It took me more years than I'd like to admit to develop a system that allows our homeowners to make all their selections before we begin building their home. When I share this with builder friends of mine around the country who build at price points much lower than ours, they cannot believe that we're able to get all of our selections done in advance.

Here's how our system works: I've assembled an interior design team, which includes an architect, an interior architect, and interior designers. We get the plans and the interior architecture detailing completed first, and that allows our interior designers to meet with the homeowners in what we call a Concept Meeting. The Concept Meeting takes between three and four hours, and is designed to do just what it says—conceptualize what the home will look and feel like in its completed form.

Next we hold a series of meetings with the interior design team, which includes but is not limited to, visits to furniture and accessory showrooms and other fact-gathering informational meetings. Our homeowners are invited back six to eight weeks after the initial Concept Meeting for a catered light lunch meeting, where our interior design team does an entire house presentation. This includes colors, fabrics, furnishings, hard surfaces, appliances, lighting fixtures, plumbing fixtures, ceiling details, door profiles, and paint colors—everything down to the last minute details. The only thing our homeowners need to select after that is their toothbrush.

Is this a lot of work? Yes, but we do it because I learned from past experience the hard way and everyone benefits from it. After years of frustration, I've developed a presentation that works; it is a highlight event. The presentation takes between two and four hours and our interior design team goes through every detail that will be included in their new home. It's all done using a PowerPoint presentation with colors, fabrics and furniture styles presented room by room, and wall by wall, so our homeowners can get a sense of how the whole house will look and feel. Our conference room is filled with colors, textures, fabrics, and finishes that will comprise their entire home. Before I did this, I noticed that most homeowners were not able to visualize how the colors and textures would flow from room to room. They had no way of experiencing the feeling they'd have when they were surrounded by the coordinated selections they were making. This presentation gives them that experience.

If you've ever built a home, you know it can be very frustrating and time consuming without the help of an interior designer. You take your paint color chip along with your carpet sample, and drive across town trying to match your tile with a designer deco piece for your shower. Then you take those samples to the granite supply yard, drive back across town to a cabinet supply house, and swing over to the lighting store. Along the way you're bombarded with a plethora of opinions from all of the people working at all of those stores. It's not only confusing–it's exhausting!

Our design presentation is like my wife and me shopping at Nordstrom's. A professional clothier helps me select my shoes, socks, suit, shirt, and tie -the entire outfit. As I try it on and look at myself in the mirror, I glance at my wife. She gives me that affirming look that means *that is the outfit for you.* Then I ask her, "Are you sure this teal shirt goes with this tie and this suit and this belt?" Again, she gives me that affirming nod with an emphatic, "Yes."

I stand there proud as a peacock because this suit is really to my liking. It has helped tremendously to have a professional (the clothier) and my wife, affirm that I'm making a good decision—*my* decision.

That's what our design team does for our homeowners. The team takes their preferences, creates a magnificent design for their home, and affirms that they're making great choices. When we build homes, ninety percent of our homeowners will sign off on the entire selection process without making any changes. We put no pressure on them to sign off right away, but I've learned that people hire us because they know we're experienced professionals.

As they leave the event, homeowners receive a three-ring binder with all their color selections, fabrics, paint colors, cabinet chips, color prints of their lighting choices, plumbing, hardware, and furniture. This binder allows them to review their choices, look forward to their beautiful home, and enjoy the building process. They don't have to leave work early or schedule meetings at the noon hour to run across town and look at carpet samples. It's all done in advance and our homeowners love it.

If your builder does not have an in-house design team to help you in your selection process, then at least retain a competent interior designer who complements your style and, as difficult as it may be to make your selections before you begin your home, do it—for your benefit, your spouse's benefit, your family's benefit, and for the benefit of the builder's and your relationship.

If you do, you'll find greater enjoyment in the building process, and your builder can build your new home more effectively.

DAVE'S ADVICE

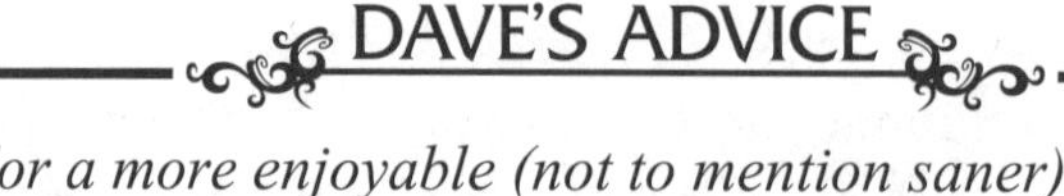

For a more enjoyable (not to mention saner) building experience, make all your selections before construction begins.

23 Will I Go Over Budget?

If you're like most people, yes, you will probably go over your budget.

Occasionally my wife will call and ask me to pick up milk and bananas from the grocery store on my way home from work. As I get out of the car, I'm thinking I'll be in and out in five minutes.

Twenty minutes later, I'm standing in the checkout line, my arms filled with items I never intended to buy. The watermelon was on sale and I know my children would enjoy it for a snack tonight after we swim in the pool. The blueberries were a two-for-one special; how could I pass up that? Seconds later, I notice fresh cut flowers out of the corner of my eye, and I know how much my wife loves flowers. Down aisle five, I see my favorite chips. I'm a little hungry since I didn't eat lunch, so I grab a bag of chips with my left hand.

Now I'm standing in line with my arms and fingers filled with items wondering what I came here for in the first place. Oh yes, a gallon of milk and bananas! I have to set the items down and run back to the dairy department because I forgot to get the milk! Does any of this sound familiar? Most likely you've been there before. The same thing can happen with the new homebuilding process if you're not careful.

As you shop for the latest and greatest appliances, you may find it enticing to upgrade. After all, it's only another $840. And the carpet is only $2.00 more per square yard than your budget planned. Then you decide that you really need nicer plumbing fixtures, upgraded light fixtures, and so on. I've found that it doesn't matter how high the budget is we help establish for our homeowners. If you're not careful, you will quickly exceed your budget. If you have an appropriate electrical fixture allowance of $25,000, for example, and I increase it to $35,000, it's still easy to overspend.

Here's what I know: Even if we bumped up every budget by 30 percent, most people will spend more than their budget allows.

DAVE'S ADVICE

Hire the right builder, establish a realistic workable budget, and then determine if you are, or are not, going to stay within that budget.

24 Don't Buy That Lot!

Call Your Builder First

I often hear from prospective homeowners who say they are almost ready to talk to a builder. They've been looking at lots and have narrowed their choice down to two or three options. As soon as they purchase their land, they want to make an appointment with the builder to discuss the home they want to build on it.

Instead of buying a lot first, we advise people to first talk to a builder. Request a meeting with your builder to look at the lots before you buy one. If you've never purchased a lot (and most people haven't), you could be walking into serious issues that you're unprepared to handle. Before you buy a lot, there are several things to consider. If you purchase land based primarily on price, you may discover unexpected additional costs and issues later on. Here are just a few items you need to know:

- Fill dirt may be needed that may cost thousands of dollars.

- Often it's necessary (especially on lakefront lots) to do soil testing to ensure that the soils are suitable to hold the home in place.

- Testing for the suitability of a septic system (if applicable) may need to be conducted. Often, local health departments will increase their requirements for septic systems. Sometimes they require a mounded septic system and that can be an unexpected additional cost. The homeowners may look at the lots adjacent to theirs and assume that the septic system can be placed below ground, only to find that a raised mound system is now required by the local municipality, and that can be very unsightly in the front yard.

- Local zoning requirements may have restrictions regarding the front, side, and rear setbacks that may be more limiting than a buyer realizes.

- Some municipalities have lot area coverage ratios. That means restrictions are placed on how large a home you can build on the lot. In some cases, there may be existing proposals to restrict lot coverage ratios. If you unknowingly have outdated information, you may be designing a home that is too large for your particular lot. You can spend six to twelve months designing a home, only to discover that local municipality restrictions preclude the building of the home you just designed, and that may mean thousands of dollars of expenses and many months of wasted effort.

Of course, this is just a short list of issues to consider before you buy a lot. I suggest you meet with an experienced, professional builder before you sign on the dotted line to purchase your lot. Obtaining current, accurate information allows you to make an informed buying decision.

Talk to your builder first; buy the lot second.

25 How Many Bids Should I Get For My New Home?

Recently I met with Ross at a lot where he was considering building a new home. About 25 minutes into the meeting, I noticed he was distracted by a car that had pulled up to the property.

I asked Ross, "Are you expecting someone?" He told me he was meeting with *six* builders that day and requesting bids from all of them. As I wrapped up the final few minutes of our meeting, Ross asked if I could give him a bid on his new home. The plans weren't complete; there were a lot of items that needed to be corrected (the specifications, including the finishes, tile, cabinetry, countertops, etc., weren't even defined), but he still wanted a bid from me.

I followed up with a letter to Ross affirming that I thought it was a good idea for him to *interview* six builders, but from my experience getting six *bids* was counterproductive. When all the bids came back, there would be no common basis for him to compare what was or was not included in the bids. In other words, he would be comparing apples to oranges to potatoes to carrots to tomatoes, and this would only add to the confusion of an already involved process. I suggested that Ross narrow his scope down to one or two builders that he connected with and felt he could trust, like, and respect, and then put focused energy in working through the details of his new home.

A few days later at a local builder's meeting, I ran into the builder who had arrived after me at the lot that day with Ross. I asked him about his meeting with Ross and he said the project was too crowded for him. Ross had also asked this builder for a bid and he was going to pass.

I'm not advocating you only interview one or two builders, but I am suggesting you narrow your search down to one or two builders. When I know there are many builders competing for the same job (especially if the job doesn't have a clear, defined focus), I decide to put my limited time and energy where I can be most effective. That's with people who have narrowed their scope, have a reasonably good idea of what they want, and who value what I value. Then we can put more concentrated effort in addressing their needs and concerns.

In our company, we very seldom bid our jobs against other builders because from the onset we've made a connection, discussed our fee structure, and demonstrated our competency. As a result, my company can give our homeowners very focused service. We can't do this with every prospective homeowner that asks us to bid on their home, because it takes a tremendous amount of energy, time, and focus to execute the process of home building with excellence.

Interview builders first and select your builder based on trust and respect, not necessarily the lowest bid.

26 Excellence or Perfection?

A prospective homeowner once asked me, "Will my home be perfect?" I told him from my experience in life I've found two kinds of perfectionists: one who asks for perfection, but realizes life is not perfect, and is very pleased with 95 to 98 percent. The other type expects perfection and is *never* happy, no matter how well a job is done.

If you are the latter, *please don't build a custom home.* Life is too short and too wonderful to take two years—or more—out of your life only to be disappointed in people and processes that are not perfect.

People build custom homes. *People* are not perfect.

It doesn't mean you can't or shouldn't expect *excellence.* Here are a few ways that excellence differs from perfection:

- Excellence is taking people and materials that are imperfect, and executing a process to its very highest level.
- Excellence is a home that's done on time.
- Excellence is what happens when something goes wrong and it's quickly recognized and corrected. (Trust me, when you build a custom home, things will go wrong.)
- Excellence is when your builder acknowledges his mistake, asks for forgiveness, and corrects it timely without pointing fingers.
- Excellence is clear communication.
- Excellence is a quality home built with straight walls, functioning doors, and overall good quality.
- Excellence is moving into your home and having your dishwasher, garbage disposal, and gas grill all operational on move-in day.

- Excellence is a phone call from your builder if something unexpected comes up and the schedule needs to be modified.

- Excellence is being asked by your builder on a regular basis, "Is there anything else we can do for you?"

- Excellence is having a homeowner so pleased that when the topic of home building comes up, he says, "Let me tell you about my builder."

Planning, designing, and building a new custom home for you and your family can be an exciting, rewarding experience if you select a competent builder who is committed to *excellence* and you have a clear understanding of each other's expectations. Home building can be a miserable experience if you are a perfectionist who is unhappy even if your builder builds to excellent standards.

If you require perfection, don't build a custom home.

27 The Startling Step Most New Homeowners Fail to Take

Obtaining feedback from previous homeowners is critical to selecting the right builder for your new custom home. In our marketing material and on our website, we list our previous homeowners so that new prospective homeowners can call any of them and ask how we performed with their project. I encourage our prospective homeowners to ask previous homeowners what we did very well and to list some of our weaknesses.

Of course, I realize we can't be all things to all people. We definitely have some strong points, but we also have limitations. If a homeowner is looking for me personally to give them a daily update on their accounting, then I'm not the right builder for them! I readily admit I only took one accounting class in college. Accounting is not one of my personal strengths. However, if someone is looking for a builder who's committed to excellence and integrity, has a great design background, and builds quality custom homes, then I am absolutely the builder for their project.

Every builder has their own unique background and unique strengths and limitations. I want people to recognize that and learn what's important. Look for the things that matter: experience, excellence, integrity.

I never try to be someone I'm not. That's why I encourage people to ask about our company's strengths and limitations. The last thing I want is to be in the middle or end of construction and have the homeowner and me looking at each other wondering what went wrong.

Here are a few questions to ask a builder's previous homeowners:

- Did your builder finish your home on time? If not, why not?
- Did your home come in on budget?
- Was there ever a time you felt your builder was being untruthful?
- Did the builder communicate to you clearly if there were any additional charges that you would incur?

- What are your builder's best qualities?
- What are some of the builder's weaknesses and limitations?
- And the most important question: Would you have your builder build for you again the next time?

Talk to the builder's previous homeowners.

Why Picking the Right Builder Is Half the Battle

28

How important is it to choose a good builder? It's #1 on the list!

You will have a lot of decisions to make as you build your custom home—selecting colors and finishes, determining size and layout, and more. But no decision will impact your homebuilding experience more than the all-important decision you'll need to make right at the start. You need to choose a great builder!

The fact is you can't do a good deal with a bad person. No matter how hard you work, you can't make a silk purse out of a sow's ear. The same is true with your builder. Here are the three most important qualities to look for in a builder for your custom home:

1. TRUST: Building a new home is probably one of the largest investments a person will ever make. If you don't believe your builder has your best interests in mind, you're talking to the wrong builder. You want a builder who works on your behalf, not someone who is only looking out for his own interests. When you call your builder, will you get a straight answer? If something goes wrong on your job, do you trust your builder to make it right? Believe me, when you build a custom home, there will always be bumps in the road no matter which builder you select. There isn't a contract written yet that will cover every possible condition that you may encounter while building your new custom home. Ask yourself: *What is your builder's intent?* If his internal compass is pointing north (with intentions to do the right thing, even when it's difficult) you're halfway there.

2. COMPETENCE: Does your builder have the ability to see and execute your home from start to finish? Is he able to walk you through the concept and design stage to produce a home that will reflect your needs, wants, and lifestyle? Does he have the right interior design team to compliment your taste and preference? Does he have the right office staff that will politely and professionally respond to your needs and questions? Does he have an on-time, competent construction team to oversee the building of your new

custom home? Does your builder have systems and procedures in place that will allow him to execute the building of your home in an excellent way? Does your builder have a warranty department to handle warranty items timely and professionally? Does your builder have a reputation and history in the community that speak well of his business?

3. FAIR PRICE: Does your builder charge a fair price? Most people start by looking at cost first. While price is certainly important, it's not nearly as critical as trust and competence. That's why I put this quality last.

If you know your builder is charging you a fair price for his services, and if you know he is trustworthy and competent, *look no further*. You have found your builder. Hire him and begin to focus on how to make your dream home become a reality.

Look for trust, competence, and fair pricing in selecting a builder to build your custom home.

29 What Kind Of Warranty Can I Expect?

An important question to ask your builder within the first few meetings is, "What kind of warranty can I expect?" You'll want to know if he provides the minimum warranty allowed by law or if his reputation and written warranty exceed your expectations.

At our company we provide a 2-10 HBW Warranty on every home we build. This warranty is a product we purchase from an independent firm to give the new owners another level of assurance that they will be taken care of if there are any problems with their new home.

The HBW Warranty provides for warranty coverage for one year on workmanship; two years on electrical, plumbing, and mechanical distribution systems; and ten years on the structure itself.

This is a value-added service that's important for our homeowners' peace of mind. But what's most important is what previous homeowners say about the warranty. My philosophy has always been that warranty is an extension of our marketing. I provide a list of our previous homeowners and encourage our prospective homeowners to call them and ask how we've done with our warranty and if there is anything my company has refused to do for them. Did we exceed expectations not only during the warranty period, but did we also show a willingness to help and correct things beyond the warranty period?

I'm not allowed to provide a written warranty that exceeds the 2-10 HBW Warranty because it would be in conflict with my warranty provider, but I can make choices to handle things that are not required of me, and for this reason we continue to get referrals over and over again.

When someone buys a Lexus and something goes wrong with their car one day after the warranty expires, there's a good chance the dealership will do whatever is necessary to keep the customer happy. After all, people who buy a Lexus have different expectations than people who buy a Hyundai or a Chevy.

People spend their hard earned money with us and we work hard to exceed their expectations. Are we perfect? No. But I believe the sign of a great company is how you handle problems. We work very hard to exceed people's expectations.

Ask your builder what kind of warranty he provides and what his philosophy is behind his warranty.

30 What About Storage?

If you are like most Americans, you accumulate stuff. Stuff takes up space, sometimes a lot of space.

Over the years, you may have accumulated things with sentimental value, seasonal items (like decorations you use once a year), or extra playthings for those wonderful visits from the grandchildren. In the homebuilding process, many people fail to take into account their need for storage, and if they do, they generally underestimate the amount of storage space they will need. Since I've moved from the Midwest (where basements are common) to Florida (where basements are rare), I've learned how to solve the storage problem.

In a new home design, people are generally willing to pay more for a home with adequate storage than a home with more finished space that's seldom used (e.g., extra bedrooms). In Florida where I build, the best storage space is generally found in a walk-in, easily accessible, partially tempered attic space. This is often located on the second story with direct access from the common area such as a game room or shared space. Adding storage space can often be accomplished in the early design stages if your builder knows this is a priority for you. The additional cost is minimal as long as your builder knows well ahead of time so he can direct the architect accordingly.

If you're not building a two-story home, a common place for storage is the attic space above the garage. Your builder can make changes to roof framing to allow for light storage and easy access above the garage. Building in a market that has demographics of an aging population, we've become sensitive to the safety of using a pull-down ladder to access this space. We've found that upgrading these access ladders to light commercial aluminum ladders has been money well spent for the safety and ease of access for our homeowners.

We've also designed storage attic spaces above garages that have their own separate set of stairs. The stairs are much safer than pull-down stairs, which are often not carpeted, and have a very simple handrail for safety. This is an additional value-added benefit not only to the homeowners but also for value at resale.

Another common storage solution is to create walk-in storage closets within the home itself. It's important to think through your storage needs as you begin to design your new custom home.

DAVE'S ADVICE

Don't forget about storage. Tell your builder how much of a priority storage is to you and look for opportunities to create inexpensive storage space.

PART II

DURING CONSTRUCTION

31 Don’t Sweat the Small Stuff

A few years ago, Richard Carlson wrote *Don’t Sweat the Small Stuff – and It’s All Small Stuff.* In his now famous book he said, "Often we allow ourselves to get all worked up about things that, upon closer examination, aren't really that big a deal..."

That’s not only good advice for life, it’s especially important during the home building process. I can assure you there's going to be a lot of small stuff during the building of your home.

People who call our office worrying about minor things make the building process much more difficult, both on the builder and on themselves. A homeowner who worries when a subcontractor is two hours late to the job, or needs to know why a two-by-four has a knot hole in it, or notices some sawdust in a corner of the living room can take the wind out of anyone's sail and cause delays.

We encourage our homeowners to let us know if there are things that truly concern them because we pride ourselves in providing a complete and pleasurable experience. However, people who view everything as a “big deal” and worry about everything (especially those who call us daily with their current worry list) are never going to be satisfied.

Jordan was someone who excelled in “sweating the small stuff” during the building of his new home. Throughout the design and contract stage of his new home, Jordan was a delight and seemed to be the perfect candidate for a successful project. We mentioned to Jordan the surveyors were scheduled for Thursday, and even though we didn’t need the survey work done for at least two weeks, we scheduled it early so it wouldn’t be a critical component in our schedule.

At 7:00 Thursday morning, it was raining buckets, and continued to rain all day. Early Friday morning, before our offices even opened, Jordan called because he was worried about the surveyor. We explained to him that because of the full day of rain the previous day, the surveyor was delayed by a day. Jordan was stressed. It was the first of many times throughout the process Jordan was “sweating the small stuff.”

A week later, when the material was dropped at his lot so construction could begin, the delivery truck got stuck due to all the recent rain. We got another call from Jordan wanting to know all the details about why there were tire ruts in his front yard.

The day the foundation man was scheduled, he was delayed because of traffic. Another phone call from Jordan. Once the foundation work was prepared, Jordan called to find out what day the inspection would occur. Once we passed inspection, Jordan wanted to know what the inspector said and why the inspection card in the permit box was signed off in black marker instead of blue ink.

After the foundation was installed, there was a bag and a half of mortar left over and two wheelbarrows of sand. Jordan called to ask what was going to be done with the leftover material.

It went on and on and on, throughout the entire job! Jordan continued to "sweat the small stuff." No amount of meetings and explanations could convince Jordan to let us do what he hired us to do. It was counterproductive to our relationship and impeded our ability to execute with excellence. We spent more time answering Jordan's questions about the small stuff than we did looking for opportunities to build his home in the most efficient and effective way possible.

At the end of the day, we all want the same thing: a beautiful, quality home. It's important to know if you're going to sweat the small stuff (and remember, as Richard Carlson said–*it's all small stuff*), it will not help you get what you really want. Let your builder do his job. We'll worry about the small stuff and the big stuff—it's what we do, and we do it well.

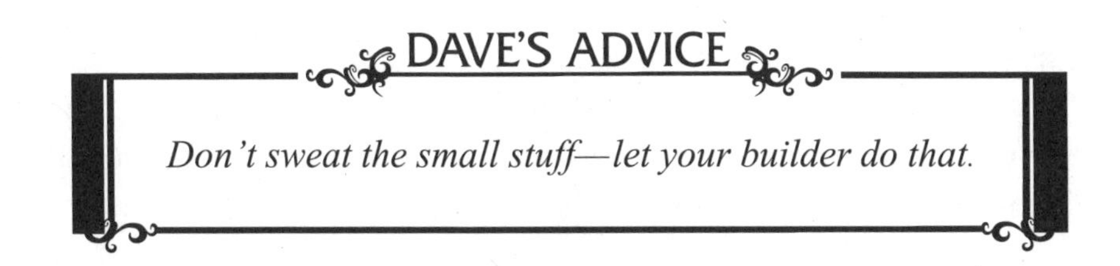

32 What Language Are You Speaking?

"I didn't buy this home at a scratch-and-dent sale. See that chip in my bathtub? I don't want it repaired. I want the entire tub torn out." Rick was clearly not happy. He obviously had definite expectations and, in his opinion, they weren't being met.

It's imperative to have a clear understanding of your expectations during the construction process if something is scratched or damaged. There are over 100,000 components that go into a new custom home, and in the process of installation something may inadvertently get scratched or damaged. I recommend you agree in writing with the builder that if he can bring the damaged item to a new quality standard, this is mutually acceptable. If you took delivery on a new car and it had a minor scratch, you wouldn't expect the dealer to replace the entire door or the entire car. Agree on standards with your builder before you begin.

I was once hired as a mediator to help resolve a conflict between a homeowner and a builder. Instead of going to a jury trial, the builder and homeowner had agreed to binding arbitration, and I was hired as the mediator. My opinion would be binding.

When I arrived at the house, I was introduced to Dr. Jones, the homeowner, who was already dressed for work in his medical scrubs. I was also introduced to Mr. Jenkins, the builder, who also arrived dressed for work in his cowboy boots, blue jeans, and a Harley Davidson t-shirt. From the start, I felt as though they were speaking different languages.

No wonder they had conflict! They came from two different worlds. The doctor was trained in exacting measures. He was trained to do things right the first time–every time. In his world, there were no second chances. A surgeon cannot tell a patient, "Oops, I forgot. I left a scalpel inside your stomach during the operation." But the builder in cowboy boots and blue jeans was thinking, "What's the big deal? The wall was put on the wrong side of the line. We can move it in ten minutes."

Years later (and thousands of dollars in attorney fees later), we had an angry homeowner and a frustrated builder; each man was looking for something the other could not provide. The surgeon was looking for a deal, and the builder was looking for the opportunity to say he built for a doctor.

In the end, I provided my written binding judgment, but neither ended up happy. The sad thing is all of this could have been avoided if they had understood what they didn't know about the other person's point of view.

I suggest you and your builder have your expectations defined and clearly written out before any construction begins. Building a new home involves so many components, and there will be things to deal with all along the way. Be clear about how your builder will handle any issues.

DAVE'S ADVICE

Define the expectations in writing before construction begins.

33

Eight Common Arguments Builders Have With Homeowners

... and How To Avoid Them

1. *Homeowner thinks: "You never finished my punch-out walk-throughlist."*

   At closing, we assemble a walk-through list. Our Construction Manager and our homeowner walk through the entire house to determine if there are any areas or items that still need attention. We schedule another appointment for two weeks after the initial walk-through to determine if there are any other items our homeowners may have found. It's important to have both of these lists in writing and signed by the homeowner and the builder. If not, the list will never end. Your builder will become frustrated when items are continually added to the list; homeowners will be frustrated because they will feel as though the builder never completed the original list. Get the list in writing and agree that if any additional items arise beyond the initial walk-through or the two-week walk-through, you will create a new, separate list.

2. *Homeowner thinks: "I didn't think adding two more windows to my $3,000,000 home would be an extra cost. After all, I'm paying $3,000,000 for this home."*

   Changes need to be clearly communicated and put in writing to protect both parties and the relationship.

3. *Builder thinks: "These homeowners have unrealistic expectations. I can never please them." Homeowner thinks: "This is a shoddy builder. I never would've hired him if I had known this."*

   Before signing a contract, both the builder and homeowner need to clearly outline their expectations. While this may take a little more time, the effort is well worth it.

4. *Builder thinks: "The homeowner doesn't have sufficient funds for changes." Homeowner thinks: "The builder didn't communicate changes clearly and in a timely manner."*

Agree in writing regarding any changes that occur after the contract is signed. I also recommend homeowners pay for changes at the time of the change, not at the end of a job.

5. *Homeowner thinks: "My builder is not taking my concerns seriously; they are falling on deaf ears."*

   Have regularly scheduled meetings with your builder to update the schedule, changes, homeowner concerns, and items that the builder needs in order to complete the home. That way, you don't have to feel like you are nagging the builder and he doesn't have to feel like construction is being halted every time he turns around.

6. *The homeowner says he spoke to the subcontractor, and the subcontractor said he could do something for the owner without the builder's knowledge.*

   All communication must be communicated through the builder or Construction Manager who is running the job. This will avoid "he said/she said" misunderstandings.

7. *The homeowner is speaking to everyone but the builder on matters related to the home or its construction.*

   Open and honest communication with the builder or Construction Manager is vital, not just with anyone who will listen. Let the builder do what you hired him to do.

8. *The homeowner is continually second guessing the builder and the decisions he is making.*

   Take time at the start of the project to interview and gain a high level of trust with the builder and his abilities. Also, speak to previous homeowners about their homebuilding experience with this builder.

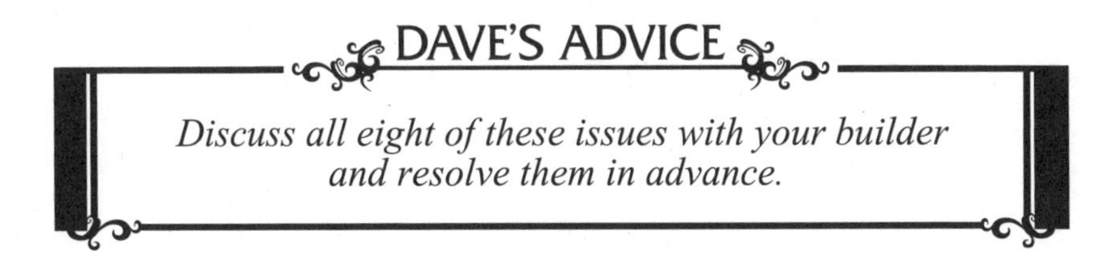

34 How Long Does it Take to Build a New Custom Home?

The amount of time it takes to build a new custom home depends on the size, complexity, and location (what part of the country you are building in).

We build homes in Central Florida and I'm often asked how long it takes to build a new custom home. The time it takes just to complete the architectural plans depends on how timely you can make decisions and your availability to meet with your architect and builder for design meetings.

On average, developing a new set of plans that is ready for permitting can take between three and six months. Add one more month for permitting. For a home that's approximately 3,000 square feet, anticipate about an eight-to-ten month construction time, provided the market isn't overly busy and there is a timely response from vendors and subcontractors. If you're building a 6,000 square foot home, anticipate a construction time of twelve to thirteen months.

For a 10,000 square foot home, add two months to the design time and another six to eight months of construction time. These estimates assume normal market conditions, which allow for a timely response from vendors and subcontractors.

It's important to understand the realistic timeline from your builder and the things that may delay a timely completion process. Beware of a builder who promises an overly idealistic timeline just to get the contract. In the end, you'll be stuck with the reality timeline.

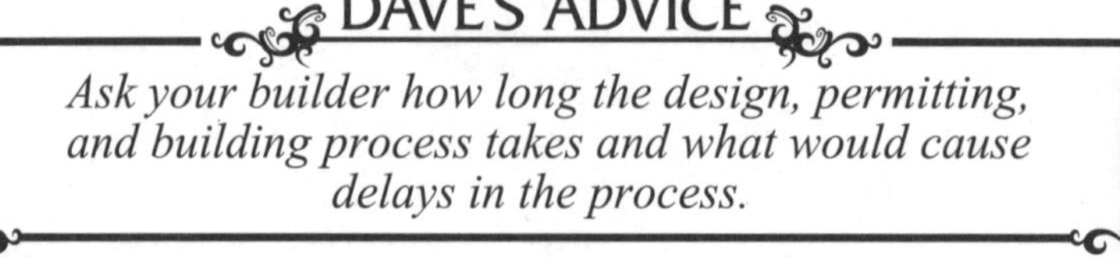

DAVE'S ADVICE

Ask your builder how long the design, permitting, and building process takes and what would cause delays in the process.

35 Understanding Two Worlds: Yours and Your Builder's

I once built a custom home for a world-class race car driver. The year I built Joe's home he was ranked number one in the world and won the driving circuit for that year.

Early on in the construction process, I couldn't understand why it was so unsettling to him if things didn't go quite as we had planned. If a subcontractor showed up a day late—even with a legitimate reason—Joe was upset. I was feeling a bit of a disconnect with him.

That year, Joe was kind enough to give me complimentary tickets to a big race. My friends and I watched the race from noon until 8:00 p.m. It was a fun day, filled with all the excitement of the race, the crowds, and the cars.

I went home, went to bed, woke up the next morning, and turned the television on to see the latest report on this exciting 24-hour race. At that moment, our seeming disconnect suddenly made sense. After 21 grueling hours of racing, Joe was leading the race by a mere ten seconds!

It dawned on me that day that in Joe's world, ten seconds was everything. The precision required to be a world-class race car driver was very different than the precision required to build a home. To Joe, having a subcontractor show up a day late was incomprehensible. In this 24-hour race, Joe was part of a three-driver rotation within the 24-hour period. If, during the driver exchange, one driver bumps his knee on the door and loses four or five seconds in the transition, it can cost him the race.

The exacting standards we have in the construction industry were just *different* than the exacting standards of race car driving. Until that moment, I didn't understand Joe's world.

Before you enter into an agreement with your home builder, the two of you should seek to understand each other's worlds. If I had done this with Joe, it would have saved us unnecessary turmoil in the building process.

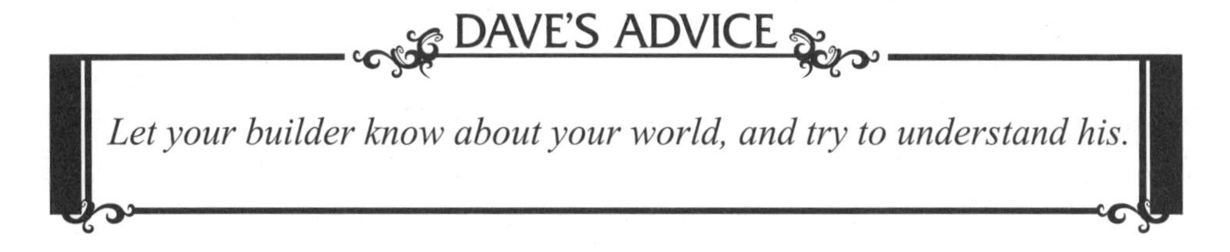

DAVE'S ADVICE

Let your builder know about your world, and try to understand his.

36 Should I Be Afraid of a Change Order?

No, you shouldn't be afraid of a Change Order–*if it's done right.*

What's a Change Order? Basically, it's a document that's used during the homebuilding process to let the builder know what you want to change from the original design specifications. For example, you may ask for a change because you want a different selection; you've changed your mind about countertops. Originally you wanted tile, but now you want granite. You can also use a Change Order to provide for a more functional use of space.

Your builder may also initiate a Change Order because material is no longer available, shipment delays have occurred, or a new and better product has become available.

The change listed on the Change Order must be described clearly with a fully researched price. Your signature will be required. A change may or may not affect the contract price of your home or the delivery date of the home. But, here's what really matters: You need to know! A Change Order does that.

So even if you initiate a Change Order for upgraded granite countertops and you agree to pay the extra cost, you still need to understand that shipping delays may bump the construction schedule back three weeks. Changes are possible, but they have consequences. The positive thing about a Change Order is everything is written down, you sign it, we sign it, and we're back in business. Nobody is going to hear, "but I THOUGHT you said..." Changes are clear and documented.

Some homeowners may be afraid of a Change Order because most builders don't process them well. Can you imagine the chaos that would occur if we just proceeded with phoned-in instructions? I never want to sit at a table with one of my Construction Managers on one side, and the homeowner on the other and hear each saying, "but I'm *sure* he said..."

Whether the Change Order is a large or small item, it always creates a wrinkle in the construction process. Some section of work has to stop until we know if we need to change direction. We do research, make phone calls, and solicit new bids, and our compensation for this additional work comes from a percentage mark-up on the dollar value of the change. This is commonly called a builder's margin, and it's generally 25%.

A Change Order may be due to a product being discontinued or there may be extensive delays in delivery of the original selection. Occasionally there are shortages of material, and the builder may propose using another material in order to keep on schedule. This action would prompt a Change Order for you to review and sign. Remember, the Change Order protects you—it's YOUR home, and you remain in the driver's seat.

DAVE'S ADVICE

Minimize the number of Change Orders by hiring a builder who communicates clearly upfront, before you start building. When a Change Order occurs, make certain a paper trail follows for your protection and the protection of your relationship.

37 Why Do I Have to Pay a Builder's Margin on a Change Order?

As a home builder, I often hear, "Don't builders make enough profit so they can just include the changes as we go along? After all, we're building a custom home and we're entitled to make changes. Otherwise, we would've bought a home from a production builder."

While this seems like a valid point from a buyer's perspective, most people would be surprised to find out that builders don't make as much money as they think. I always encourage homeowners to work diligently on the front end (before construction begins) so they can keep Change Orders to a minimum. Changes can happen during the construction of a custom home, but you need to be aware of how the process works, the consequences of a change (additional time and cost), and understand the builder's margin.

I've built homes with as little as four changes and I've built homes with more than 200 changes. It's difficult to gauge how many changes a homeowner may make after signing the contract. Most people start out saying they love their plans, and they love everything they've picked out. Yet once construction starts, we may receive nine phone calls requesting 20 changes during the first week alone!

Changes involve a lot of energy and a lot of the builder's team's time. In order to effectively process and execute changes to a custom home, a builder needs to be fairly compensated. The last thing that you want is to have your builder wincing when he looks at his mobile phone when you call. I'm not saying that a builder should be able to take a trip to Hawaii because of a single Change Order fee he earns while building your home, but a fair and equitable fee that encourages your builder to work on your behalf for an excellent result is vital to the homebuilding process.

We don't encourage or discourage our homeowners to make changes. Being available to give professional counsel when questions about changes occur is our duty and responsibility. We're motivated to give our homeowners a great experience.

I once built a home for a bank president. John was a great client and a great guy. Our contract agreement outlined that he would provide his own refrigerator. Near the end of construction, John asked if he could use my supplier to purchase the refrigerator at my cost. Since we had a great relationship, and my focus was on serving, I allowed him to select the refrigerator.

I had it delivered, and only requested reimbursement of the cost of the appliance, without charging him a builder's margin.

In my original agreement, one of the items that this homeowner valued was an extended warranty, which I provided. Approximately fourteen months after John moved into his new home, the seal on the refrigerator failed and water leaked onto the hardwood floor, causing the floor to warp. When I contacted the appliance company, they indicated that the refrigerator was out of warranty, and since the home was still in warranty (because of the extended warranty period I had provided), I had a problem. Not only did I have to pay to repair their refrigerator, but I also had to pay for repairing the floor and sanding and finishing the entire floor in his home because the new finish didn't match the original finish in the rest of the home.

I learned a valuable lesson from that experience. If changes occur that deviate from the original contract agreement, then a builder's margin must be charged to compensate fairly for the time, energy—and risk—associated with items that may need to be addressed at a later date.

Agree in advance what the builder's margin will be on Change Orders. Decide what you want before construction begins, and if a change occurs, you'll have a mechanism that provides for a win/win situation.

38 How To Make Your Builder Love You

I've built hundreds of homes in my career, and I've worked with some really great people and some very difficult people. Our most successful projects have been the result of developing great relationships. In addition to providing a luxury custom home, many of these homeowners have become my lifelong friends; some are now my very best friends.

While not every homeowner will form a lifelong friendship with their builder, here are some tips on how to have a successful project and make your builder love you:

- TELL THE BUILDER WHAT HE IS DOING RIGHT. I once worked with a homeowner who found something good to say every time we spoke. Yes–every time. Nick didn't do this in a patronizing way; he gave me and my team genuine compliments. He looked for and commented on the positive aspects in our relationship and the services we were providing. Nick also shared his concerns with me. I would do anything for Nick. I'd run through a wall for him and would do so to this very day, more than five years after completing his home.

- CLEARLY COMMUNICATE PROBLEM AREAS. If something is bothering you about your new home construction process, clearly communicate what concerns you without anger or a condescending attitude. Give your builder an opportunity to make it right. A demonstration of a great builder is how well he handles problems.

- CLEARLY COMMUNICATE YOUR EXPECTATIONS. Be forthright and share with your builder what you really value and tell him what is important to you. If you're clear, you'll probably get what you want.

- LET YOUR BUILDER DO HIS JOB. We've been hired by homeowners in the past who seem to be agreeable, only to find out later they wanted to control the entire process and hover over us. Behavior like that makes our entire team reluctant to make a decision for them.

- THINK WIN-WIN. Builders are regular people, just like your next door neighbors. They generally don't make as much money as people imagine, and most builders really want to do a good job. Work towards amicable solutions.

- REMEMBER TO SAY THANK YOU. Builders are people too and everyone likes to hear a thank you for a job well done.

Follow these helpful hints and your builder will love you.

39 Why You Shouldn't Use Friends as Subcontractors

I have a friend (or a brother-in-law) who is willing to do the tile work in my new home for a discounted rate. I'd like to use him and save some money."

Early in my career as a custom home builder, I learned it's not a good idea for homeowners to use friends or relatives for subcontracting work. I made the mistake of allowing people to have a hand in some aspect of the building process and most of the time it was a disaster for me and for the homeowner. Actually, it only worked out well once or twice in about 25 homes where I allowed it.

In one memorable example, I allowed Scott to use his friend's custom cabinetry shop. Scott told me he knew the friend well and had spoken to another friend who was pleased with the cabinetry from this particular company. My gut told me not to allow it, but against my better judgment I agreed to let Scott use his friend's cabinet company.

We were promised the cabinets by a specified date, but the cabinets were not installed on time, which caused a two-month delay in our construction schedule. Scott and his family moved in without any cabinet doors on the face of the cabinetry in their entire home and then waited an additional two months to have their job complete. That doesn't even take into account the lost momentum, additional cost in interest carry, overhead, and other expenses as a result of the delay from the cabinet shop.

When you hire a builder, I strongly encourage you to allow your builder to do what he does best. When you visit a dentist, you expect him or her to be trained and equipped to do the job efficiently and effectively. You wouldn't think of bringing your own tools or materials, hand them to your dentist, and ask if you could save money on a filling because your neighbor or friend is in the dental supply business. You wouldn't take a steak to a restaurant and ask them to cook it for you.

The principle is no different in home building. If you use friends or your own business contacts, it will disrupt the smooth flow of work and communication of what an experienced builder does best. Trust your builder. He has vendors, subcontractors, and a labor force already in place that he works with on a regular basis.

DAVE'S ADVICE

Hire a competent builder to do the building;
save your friends for the housewarming party.

40

The Top Eleven Mistakes Made by Homeowners

- *Purchasing a lot that is "affordable."*

  Solution: Remember: location, location, location. Purchase the most expensive, valuable lot you can manage, even if it means waiting on some finishes or amenities in your home. (See chapter 6 for more on this topic.)

- *Trying to build a custom home without a professional builder.* Building a custom home is more complex than most people realize. It takes skilled professionals years to learn the business and even then, changes in the industry, materials, and codes make it difficult to keep up.

  Solution: Find a competent builder you can trust. Negotiate a reasonable fee for his services and hire him (see chapter 3).

- *Purchasing a ready-made plan thinking it will save you money.* Building someone else's design or dream (especially one that was designed for someone in another city and state) may not be the wisest choice.

  Solution: Purchase a ready-made plan only if your lot is standard, and you don't need to modify the plan (see chapter 12).

- *Choosing a builder primarily because of price.* The expression "you get what you pay for" applies to the home building process. If you've heard horror stories about people's experience with their builder, it usually can be attributed to someone trying to get a deal.

  Solution: Your home is a major investment. Make an informed, purposeful, thoughtful decision and don't be lured by the lowest bid (see chapter 17).

- *Biting off more than you can chew.* In an appreciating market, the rise in value can cover this mistake, but in a flat or declining market, it can be disastrous.

  Solution: Know what you can afford and stick to your budget (see chapter 23).

- *Hiring a builder when your gut instinct tells you not to.*

Solution: After careful research and comparing builders, go with your instinct, not the discounted price (see chapters 4 and 28).

- *Making choices for your home that only you love, but everyone else hates.*

  Solution: Get good counsel from your builder, architect, interior designer, and real estate professional before you make your decisions (see chapter 14).

- *Expecting workers to be on your job every day from 7:00 A.M. until 4:00 P.M.*

  Solution: Recognize that some days, no work is scheduled at all because inspections may be taking place or rain has caused a change in the schedule (see chapter 31).

- *Underestimating the importance of making all selections before construction.*

  Solution: Make all selections prior to construction and enjoy the building process (see chapter 22).

- *Homeowners giving direction to subcontractors on the job.*

  Solution: Communicate only with the Construction Manager or builder. The Construction Manager is the only person on the job who has *all* the information related to your project. Subcontractors have only one piece of the puzzle. You can visit the job site during scheduled appointments with the Construction Manager, who can answer your questions and explain what you will be seeing.

- *Not understanding the "Change Order" process.*

  Solution: Discuss the builder's Change Order process with him and be sure you are clear with how it works. Cooperating fully with this process will go a long way toward your enjoyment of the whole project (see chapter 36).

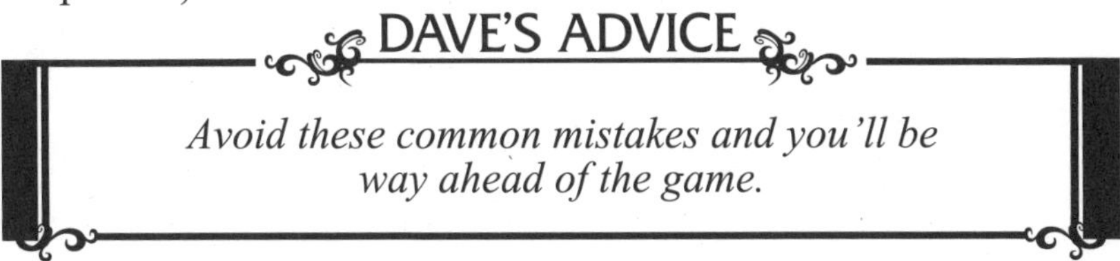

41
Should I Hire An Independent Building Inspector?

Sometimes homeowners choose to hire an independent building inspector during the building process of their new home, but that decision can be a double-edged sword.

On one hand, if the building inspector's intent is to genuinely help the process by effectively communicating what he observes, it can really aid in the process of completing a new home. On the other hand, a building inspector who tries to justify his fee by searching for insignificant things can add confusion and even create an adversarial relationship between the builder and the homeowner.

We don't discourage our homeowners from hiring an independent building inspector. It's important to us for them to be fully comfortable with the building process. After all, they are investing a lot of money in their new home. If the homeowners can clearly define, in positive terms, what they want the inspector to do, the inspector will know we're not in an adversarial position. We will all be on the same page, working together to create a beautiful home.

Most building inspectors charge between $500 and $1,000 per inspection. However, much of what is noted by a private building inspector is already being handled by the builder or the required local, county, and municipal building inspectors at code inspections. Hiring the private building inspector may be additional cost without any real value added to the homeowner.

At the *completion* of our homes, we often hire an independent building inspector. We do this for two reasons: one is to have an independent, professional set of eyes check out all aspects of the home, including mechanical components such as plumbing, air conditioning, electrical, appliances, dual garage door openers, etc. They also do a thorough inspection of roofs, windows, and doors to ensure no leaks will occur, and they make sure there are no code violations that may have been inadvertently overlooked by a local building official.

The second reason we hire an inspector is so the new owners will be certain of the home's excellence. Our homeowners may not even realize that we pay for our own independent inspection of their home. The inspection report, which is sent to us electronically, contains digital photos and clear notations of items that need to be addressed.

As a company, we've chosen an independent building inspector who is very picky; this helps us perform at our highest level.

If you hire an independent inspector, make certain he is helping, not hindering, the building process.

PART III

HELPFUL CHECKLISTS

CHECK LIST #1

What To Do Before You Hire a Builder

This book would not be complete without a *"Helpful Checklists"* section. Each checklist is designed to make it easy for you to keep on track throughout the entire home building process. You may want to review the lists and check off the items as you go through each phase. It can be helpful to take them with you to meetings with your builder or to the building site.

Your builder may have his own checklists for various stages of the design and build process. While they may differ in some details, they will most likely be quite similar.

These checklists have been helpful and valuable to our homeowners and to our company. They allow us to verify items that are significant in the design and construction process and make the process more efficient. The following checklists are not exhaustive; they simply highlight some of the "must do's" while your new home is being built. We hope these tools will help you in your home building experience.

- ☐ 1. Look at some of the builder's homes currently under construction.
- ☐ 2. Look at some of the builder's completed homes.
- ☐ 3. Have the builder provide a list of previous homeowners.
- ☐ 4. Call two or three previous homeowners this builder has worked for and ask them key questions (see Checklist #3).
- ☐ 5. Become comfortable with the contract documents before signing.
- ☐ 6. Decide on your home building budget.

CHECK LIST #2

Top Questions To Ask a Potential Builder

43

- ☐ 1. Why should I hire you?
- ☐ 2. What is your fee structure?
- ☐ 3. How do your fees compare to other builders' fees?
- ☐ 4. What are your weaknesses?
- ☐ 5. What makes you different from other builders in this market?
- ☐ 6. What type of warranty do you provide and what is your philosophy on warranty?
- ☐ 7. Share with me your worst building experience with a homeowner and what you learned from it.
- ☐ 8. How long have you been building?
- ☐ 9. What is your education?
- ☐ 10. How many custom homes have you built?
- ☐ 11. How many homes will you have under construction at the time my home will be built?
- ☐ 12. How many homes do you build per year?
- ☐ 13. How do you handle changes?
- ☐ 14. How many Change Orders would you consider average in building a home?
- ☐ 15. Do you supervise the building yourself, or do you have a site supervisor?
- ☐ 16. Can I meet the person who will be running my job?
- ☐ 17. How often do you come to the job?
- ☐ 18. What work do you do with your own crews, and what work do you subcontract out?
- ☐ 19. Do you have contracts with your subcontractors? Can I see a sample?
- ☐ 20. Can you provide us with a bank reference?
- ☐ 21. Can you provide us with a copy of your insurance certificate?

CHECK LIST #3

Top Questions To Ask Your Builder's Previous Homeowners

- ☐ 1. Why did you select this builder?
- ☐ 2. Did your builder demonstrate character and integrity during the time you've known him?
- ☐ 3. Did you feel your builder had your best interests in mind?
- ☐ 4. How well did your builder communicate Change Orders to you?
- ☐ 5. Did your builder treat you in a respectful, honorable way, and did he communicate clearly with you?
- ☐ 6. What was the worst thing that happened during your building process?
- ☐ 7. What would you do differently if you had to build your home all over again?
- ☐ 8. Was your home built on time?
- ☐ 9. Was your home completed on budget?
- ☐ 10. Was there ever a time you felt your builder was being untruthful?
- ☐ 11. Did your builder exceed your expectations?
- ☐ 12. What were your builder's best qualities?
- ☐ 13. What were some of the builder's limitations and weaknesses?
- ☐ 14. Would you use this builder again? (The most important question of all!)

CHECK LIST #4

Before You Begin Construction

Review the following items with your builder at the job site before construction begins. This is not intended to be an exhaustive list; it includes the main things to check before you begin. Ask your builder to explain any items not clear to you.

- ☐ 1. Verify with your builder the height of the finished floor and all porches.
- ☐ 2. Verify with your builder the type of material used on all porches.
- ☐ 3. Verify with your builder all floor outlet locations.
- ☐ 4. Verify with your builder any slab/recessed areas for hardwood application.
- ☐ 5. Verify with your builder all hose bib locations.
- ☐ 6. Verify with your builder electrical meter locations.
- ☐ 7. Verify with your builder layout and location of driveway and sidewalks.
- ☐ 8. Verify with your builder well and septic locations (if applicable).
- ☐ 9. Verify with your builder pool equipment location (if applicable).
- ☐ 10. Verify air conditioning locations. Is the unit too close to the master bedroom window?
- ☐ 11. Verify lot drainage conditions.

CHECK LIST #5

Before Your Builder Hangs Drywall

46

Review the following conditions at the job site with your builder prior to hanging drywall:

- ☐ 1. Spot check room dimensions with your builder.
- ☐ 2. Verify with your builder that all closets are in place. These are easily overlooked by the framing contractor.
- ☐ 3. Review all door swings with your builder with electrical switch placement in mind.
- ☐ 4. Verify with your builder location and quantity of all electrical outlets, switches, cable TV, computer, etc.
- ☐ 5. Verify with your builder placement of breakfast nook ceiling fixture placement.
- ☐ 6. Verify any additional electrical outlet needs with your builder such as sockets, switch lights or holiday lights, receptacles, or convenient overnight mobile phone charging locations.
- ☐ 7. Review with your builder placement of master shower valves for easy reach to avoid scalding.
- ☐ 8. Verify cabinet layout and sink locations with your builder.

CHECK LIST #6

3 Weeks Prior to Closing

- ☐ 1. Call your insurance agent to put homeowner's insurance in effect. (If there are current hurricanes, it can affect the ability to put insurance in place.)
- ☐ 2. Call your builder to verify the tentative walk-through date.
- ☐ 3. Call electric company to schedule service in your name.
- ☐ 4. Call telephone company to activate telephone service.
- ☐ 5. Call cable company for TV hook-up.
- ☐ 6. Call water municipality to schedule water in your name.
- ☐ 7. Call gas company to schedule gas service in your name.
- ☐ 8. Call pool service company for pool cleaning contract.
- ☐ 9. Call lawn service company to put contract in place.

CHECK LIST #7

The No-Sweat List for Closing and Final Walk-Through

This checklist will guide you through the process of what will happen on that long-awaited day—your closing day! After many months, it is finally time to hand you the keys. One of the important closing day events will be for your builder to walk through the entire home with you. This is an opportunity for the builder to instruct you about certain aspects of your new home, and to point out locations with critical information.

A good builder will have his own checklist, but this list will let you know what to expect, and you can check off your own items as you walk through your beautiful new custom home together. It's also a great time to ask any questions.

- ☐ 1. Verify that keys work in every lock throughout the entire home.
- ☐ 2. Obtain garage door openers.
- ☐ 3. Obtain all extended warranties.
- ☐ 4. Obtain owner's manuals for all appliances.
- ☐ 5. Run the dishwasher through a cycle to be sure it works and there are no leaks or other problems.
- ☐ 6. Verify garbage disposal operation.
- ☐ 7. Verify gas hookup to all appliances.
- ☐ 8. Verify hot water heater and recirculating pump operation.
- ☐ 9. Learn location of all air conditioning filters.
- ☐ 10. Locate emergency water shut-off valve.
- ☐ 11. Obtain subcontractor emergency phone numbers and information.
- ☐ 12. Obtain all final waiver of lien notices.
- ☐ 13. Obtain the Certificate of Occupancy.

- ☐ 14. Obtain final survey.
- ☐ 15. Obtain final Change Orders calculations.
- ☐ 16. Obtain an updated list of all colors and finish selections.
- ☐ 17. Verify pool is in working order (if applicable).
- ☐ 18. Verify gas grill is working.
- ☐ 19. Obtain home warranty.
- ☐ 20. Obtain termite certificate.
- ☐ 21. Schedule a two-week walk-through follow-up date with your builder.

The Savvy Homeowner's Glossary
45 Indispensable Words Every Homeowner Should Know and Understand

1. **Adjustable Rate Mortgage (ARM) -** A mortgage interest rate that changes based on an index over time.

2. **Agreement of Sale** – A sale contract.

3. **Amortization Schedule** – A schedule showing how the monthly mortgage payment is applied to the principal, interest, and the current mortgage balance.

4. **Appraisal** – An evaluation of homes within the surrounding area to determine the market value of the property.

5. **Appreciation** – An increase in the value of property.

6. **Borrower -** The person or persons responsible for the loan, also called the mortgagor.

7. **Cash Reserve -** The cash balance a borrower has left after closing, available for the first one or two mortgage payments.

8. **Certificate of Occupancy** - A certificate issued by a governing agency stating that the building has been approved for occupancy.

9. **Closing** - Finalizing the purchase and financing documents and the disbursement of funds to all parties.

10. **Closing Costs** – All the costs associated with the loan and the purchase, not including the actual cost of the property.

11. **Commitment Letter -** Formal notification from a lender stating the terms of the loan.

12. **Contingency** - A specific condition to an agreement or contract.

13. **Credit Report** - A report of credit history used to determine an individual's credit worthiness, usually provided by a credit bureau.

14. **Debt to Income Ratio** - The percentage of one's earnings used to qualify for a mortgage.

15. **Earnest Money** - A deposit given to a seller by a prospective buyer.

16. **Easement** - A right of way given to others to access over and across the property.

17. **Equity** - The difference between the market value and the outstanding mortgage balance.

18. **Fixed Rate Mortgage** - A mortgage in which the rate of interest is fixed for the entire term of the loan.

19. **Flood Insurance** - Insurance for properties designated in flood areas by the government.

20. **Hazard Insurance** - Homeowner's insurance.

21. **Homeowner's Warranty** - Insurance that covers repairs for the home for a specified period of time.

22. **Lien** - A legal claim against a property that must be paid when the property is sold.

23. **Loan-to-Value Ratio (LTV)** - The difference between the mortgage amount and the value of the property. Example: Home Value = $1,000,000, Mortgage amount = $900,000, LTV = 90%.

24. **Lock in Rate** - A written guarantee for a specific rate of interest by lender.

25. **Mortgage Broker -** A company that matches borrowers with lenders for a fee.

26. **Mortgage Insurance -** Insurance that is provided by independent insurers that protects the lender in the event of a mortgage default.

27. **Mortgagee -** The lender.

28. **Mortgagor -** The borrower.

29. **Origination Fee -** The fee paid to a lender for processing a loan, also called points.

30. **Owner Financing -** When the seller of the property provides all or part of the financing.

31. **Plot Plan -** A map prepared by a licensed surveyor depicting the exact placement of a house on a lot.

32. **Points -** One time charge by a lender. One point is one percent of the mortgage amount.

33. **Prepayment Penalty -** Fees charged to a borrower for paying off a loan prior to the maturity date.

34. **Pre-Qualification -** Pre-determining a buyer's financial borrowing power prior to a purchase. Pre-qualifying does not guarantee loan approval.

35. **Principal** – The total loan amount borrowed or the total unpaid balance of the loan.

36. **Radon** – A radioactive gas which, if found in sufficient levels, can cause health problems.

37. **Refinancing** – Paying off an existing loan with a new loan on the same property.

38. **Settlement Sheet** – The costs payable at closing to determine the seller's net proceeds from the sale and the buyer's required net payment.

39. **Survey** – A drawing showing the legal boundaries of the property.

40. **Title** – A legal document establishing the right of ownership.

41. **Title Company** – A company that specializes in insuring the title to the property.

42. **Title Insurance** – Insurance that protects the buyer and the lender against losses arising from disputes over ownership of the property.

43. **Title Search** – A search of legal records to ensure that the seller is the legal owner of the property and that any liens or claims against the property are identified.

44. **Transfer Tax** – State or local taxes due when title to property transfers from one owner to the other.

45. **Underwriting** – The process of evaluating a loan application to determine whether or not it's acceptable to the lender.

A FINAL THOUGHT

As I mentioned in the introduction, I've been building homes since I was ten years old. I started following my father around and helped him as much as I could when he started his own home construction business. For most of my life I've seen the value a beautiful home can have to a family. For me, a house is more than just bricks and mortar. It's more than a beautiful structure with bedrooms, bathrooms, and garages. It's an intimate living space that speaks to the very heart of who you are. It's not just a house—it's a home.

In the home I built for my own family a few years ago, I remember the bedroom we created for my oldest daughter, Lauren. She was about five at the time. We made her a "Princess Room," with an elevated bed draped in soft silk flowing fabric—just like a real princess would sleep in, we told her. Underneath the elevated bed, we created a magical, safe, almost secret space, with a door on the front. Frequently my wife or I would climb into this wonderful space with her to play games, tell stories, pray together, and reflect on God's goodness. She loved her Princess Room.

At the end of the day, I always looked forward to tucking Lauren in for the night. High up in her princess bed, surrounded by the magic of silk fabric, fluffy pillows, and flowering trellises, I'd tell her bedtime stories. Sometimes I'd share stories about my own childhood adventures growing up on a dairy farm. She never grew tired of hearing the one about her daddy being chased around the field by a furious heifer and how the crazy cow bucked daddy right off his feet. I can still hear her giggles at the sound of the snorting cow and the shriek of the flying boy. I think I told her that story 300 times—just so I could hear those giggles.

As she snuggled down for the night, I'd tell her how blessed I was to have her as my little girl, and how much I loved her and her brothers and sisters. Then, before she slipped off into her own private world of princess dreams, I loved to pray with her. Like so many parents, I prayed that God would bless her and protect her. I prayed her relationship with Him would grow deep and strong, and that she would want to serve this great God her whole life. I prayed that she would live to be a blessing to others.

After bedtime prayers came the goodnight hugs and kisses. I'd step down from the platform, switch off the lights, and close the castle door for my own little princess.

Memories. For me, that's what building homes is all about–creating a beautiful and safe structure so the children, parents, family and friends can create and share memories for a lifetime. Every day I have the privilege of helping people create unique places and spaces to make memories. I get to listen as people tell me about their own families, their wants, their needs, their values and their dreams. Then I get to bring together an incredible team of talented people who work to shape and give life to those dreams.

My hope is that this book has helped you to move toward custom designing and building your own unique one-of-a-kind home. Now that you know what's involved, you can find a great builder who will make your family's home dreams come true.

Enjoy your journey.

David A. Konkol

APPENDIX I

20 Step Pre-Construction Process

Our 20 Step Pre-Construction checklist was created to assist clients of Dave Konkol Homes in understanding the initial steps required to get to the construction phase of your new home. Use this list with your builder to expedite your pre-construction process.

- ☐ **Step 1 -** Initial contact with Dave Konkol Homes, Inc. (DKH) to begin defining home building requirements.

- ☐ **Step 2 -** Delivery of DKH Home Planner to you, including **Design Outline**™ homework.

- ☐ **Step 3 -** E-mail or fax completed **Design Outline** homework.

- ☐ **Step 4 -** Dave Konkol, president, and our architect will review your completed **Design Outline** and will contact you to discuss budget, sight parameters, home orientation, design image, and answer your questions.

- ☐ **Step 5 -** Schedule first design meeting with architect at Dave Konkol Homes (typically 3 hours). Additional meetings may be scheduled during the design process depending on the scope of the project.

- ☐ **Step 6 -** Meet with Dave Konkol and our architect to review site plan, floor plan, and an elevation study.

- ☐ **Step 7 -** Meet with Dave Konkol to review the specifications of the home, the construction agreement, and warranty of your new home.

- ☐ **Step 8 -** Meet with a Dave Konkol Homes representative to sign initial contract and provide a required deposit in order to proceed.

- ☐ Step 9 - Schedule creative meetings with Dave Konkol, architect, interior architect, interior design team, lighting consultant, and interior furniture designer to establish style, color preferences, and to further clarify your home design needs (typically three hours).

- ☐ Step 10 - At the end of the meeting with interior designers, meet with our lighting specialist to further define your lighting needs (typically one hour).

- ☐ Step 11 - Meet with interior architect and interior designer to present ceiling plans and architectural details (typically 2 hours).

- ☐ Step 12 - Final construction drawings are completed.

- ☐ Step 13 - Final construction drawings are submitted for permitting.

- ☐ Step 14 - Meet with interior design team to define design direction, theme, style, colors, and textures (typically 2 hours).

- ☐ Step 15 - Interior design team will accompany you to the home furnishings store to introduce and further define your home furnishing needs (typically 4 hours).

- ☐ Step 16 - Final design presentation will be presented by our interior design team. Presentation will include exterior and interior finishes, colors, and furnishings selected for your home (typically 3 hours). Sign off on selections are made at this time.

- ☐ Step 17 - Your signature is required on all selections before the construction of your new home begins.

- ☐ Step 18 - Dave Konkol Homes will generate Change Orders as necessary and submit for your review and signature.

- ☐ Step 19 - Construction begins upon approval of building permit.

- ☐ Step 20 - Pre-construction meeting is scheduled with your personal builder.

APPENDIX II

The Design Outline™ *SAMPLE: HIS LIST*

DESIGN OUTLINE™

HIS LIST

Name: John and Mary Jones		
Address of Property: 129 Stone Drive, Maitland, FL 32751		
Lot#: #13	Cost of Lot: $400,000.	Value of Lot Today: $500,000.
Estimated budget for home excluding lot and financing costs: $1,200,000.		
Number of Bedrooms: Four		Number of Bathroms: Five
Approx. number of square feet under air: 4000-5000		
Number of Stories: One (Two)		Bays in Garage: Two (Three) Four

COMBINED LIST OF PRIORITIES

	1. Three car garage with storage		16.
	2. High tech sound system throughout house		17.
	3. Low voltage lighting package		18.
	4. Basketball hoop at driveway		19.
	5. Cool looking pool		20.
	6. Good insulation		21.
	7.		22.
	8.		23.
	9.		24.
	10.		25.
	11.		26.
	12.		27.
	13.		28.
	14.		29.
	15.		30.

The Design Outline™ *SAMPLE: HER LIST*

DESIGN OUTLINE™ HER LIST

Name: John and Mary Jones		
Address of Property: 129 Stone Drive, Maitland, FL 32751		
Lot#: #13	Cost of Lot: $400,000.	Value of Lot Today: $500,000.
Estimated budget for home excluding lot and financing costs: $1,300,000.		
Number of Bedrooms: Five		Number of Bathroms: Four & one half
Approx. number of square feet under air: 4400		
Number of Stories: One (Two)		Bays in Garage: Two / Three / (Four)

COMBINED LIST OF PRIORITIES

1. Warm homey feeling	16. Circular drive
2. Hardwood throughout	17. 12' ceiling in entry, living, and dining; 10' rest of downstairs
3. Nice kitchen with granite counter tops	18. 9' ceilings upstairs
4. Crown molding throughout	19. Single front door with sidelights
5. Traditional elevation with brick	20. Exterior accent lighting
6. Brick pavers on driveway	21. Wood/clad windows
7. Shingle roof	22. Interior shutters
8. Larger secondary bedrooms	23. Good quality carpet
9. Walk-in closets for secondary bedrooms	24. Built-ins in family room
10. Lots of storage in home	25. Fireplace in master bedroom
11. Tall baseboards	26.
12. Double wall oven, cook top, microwave, oversized refrigerator	27.
13. Two dishwashers	28.
14. Moen plumbing fixtures	29.
15. Smooth stucco	30.

The Design Outline™ *SAMPLE: HIS & HERS COMBINED LIST*

DESIGN OUTLINE™ COMBINED LIST

Name: John and Mary Jones		
Address of Property: 129 Stone Drive, Maitland, FL 32751		
Lot#: #13	Cost of Lot: $400,000.	Value of Lot Today: $500,000.
Estimated budget for home excluding lot and financing costs: $1,300,000.		
Number of Bedrooms: Five	Number of Bathroms: Four & one half	
Approx. number of square feet under air: 4400		
Number of Stories: One (Two)	Bays in Garage: Two (Three) Four	

COMBINED LIST OF PRIORITIES

1. Warm homey feeling
2. Hardwood throughout
3. Three car garage with storage
4. Nice kitchen with granite countertops
5. Crown molding throughout
6. High tech sound system throughout home
7. Low voltage lighting pacakge
8. Traditional elevation with brick
9. Brick pavers on driveway
10. Shingle roof
11. Larger secondary bedrooms
12. Walk-in closets for secondary bedrooms
13. Lots of storage in home
14. Tall baseboards
15. Double wall oven, cook top, microwave, oversized refrigerator
16. Two dishwashers
17. Moen plumbing fixtures
18. Smooth stucco
19. Circular drive
20. Basketball hoop at driveway
21. 12' ceiling in entry, living, and dining; 10' rest of downstairs
22. 9' ceilings upstairs
23. Single front door with sidelights
24. Exterior accent lighting
25. Cool looking pool
26. Wood/clad windows
27. Interior shutters
28. Good quality carpet
29. Built-ins in family room
30. Good insulation

The Design Outline™

DESIGN OUTLINE™

Name:						
Address of Property:						
Lot#:	Cost of Lot:		Value of Lot Today:			
Estimated budget for home excluding lot and financing costs:						
Number of Bedrooms:			Number of Bathroms:			
Approx. number of square feet under air:						
Number of Stories: One	Two	Bays in Garage:		Two	Three	Four

COMBINED LIST OF PRIORITIES

	1.		16.
	2.		17.
	3.		18.
	4.		19.
	5.		20.
	6.		21.
	7.		22.
	8.		23.
	9.		24.
	10.		25.
	11.		26.
	12.		27.
	13.		28.
	14.		29.
	15.		30.

FREE BONUS OFFER

Thank you for reading this book. I hope it has been helpful to you. I invite you to send comments to me at Dave.Konkol@DaveKonkolHomes.com. Your comments, feedback, and suggestions are all welcome and will be used for the next edition of this book. Please tell me:

- What was especially helpful to you?
- What information would you like to see included in the next edition?
- Send me your own stories of success and struggles. (Please note: due to the volume of email I will not be able to answer personal building or design questions.)

In return for your valuable suggestions, I'll send you an excerpt from my next book (*Your Dream Home—Room By Room*) for FREE. In this free bonus, you'll learn the insider building secrets I give to my top clients who are building luxury custom homes.

You'll read about:

- Delightful Dining Rooms
- Gratifying Garages
- Fabulous Family Rooms

You'll learn design tips that make these three rooms the best they can be.

This information is not available in any other form and will only be available for a limited time. This valuable report is priced at $14.95, but you'll receive it FREE by return email when we receive your feedback.

Thank you for reading, and thank you for your feedback. God bless you and your family.

David A. Konkol

TO ORDER BULK COPIES CONTACT:

Dave Konkol Homes, Inc.
1000 N. Maitland Avenue
Maitland, FL 32751
407-539-2938 (Office)

Dave.Konkol@DaveKonkolHomes.com

1 – 9 copies:	$24.95 each
10 – 25 copies:	$19.95 each
26 – 100 copies:	$15.95 each

FREE postage and handling on all orders.

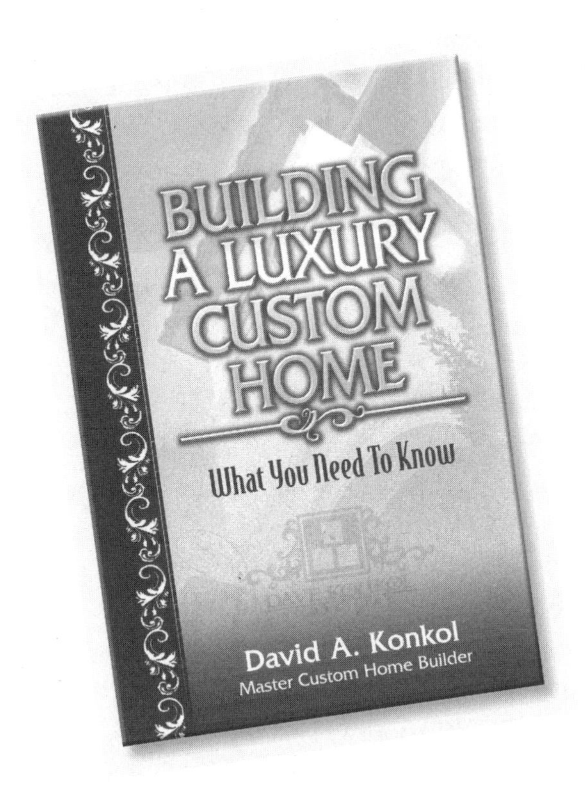

About Dave Konkol Homes, Inc.

Dave Konkol Homes, Inc. has been building beautiful, luxury, custom-designed homes throughout Central Florida since its inception in 1992.

Dave Konkol Homes participates annually in the Orlando Spring Parade of Homes, and has won numerous awards for creative design and quality construction. Dave's homes have also been showcased in the Orlando Street of Dreams Show where over 40,000 people toured his show home. Dave is an active member of the Master Custom Builder Council (MCBC), an organization comprised of 35 of Central Florida's leading custom and luxury homebuilders. Membership in the MCBC is an honor reserved for builders with a proven, time-tested reputation and is by invitation only.

Dave is also a member of the Home Builders Association (HBA) of Metro Orlando, a professional trade association, recognized as the voice and leadership of the housing and building industry in Central Florida. He is a member of the National Association of Home Builders (NAHB), headquartered in Washington, D.C. and a leader in the NAHB Builder-20 Club program. The Builder-20 Clubs are comprised of up to 20 builders from around the United States and Canada who meet twice a year to compare notes, share what works and what doesn't, and learn from each other.

Dave Konkol Homes offers the full array of services a homebuilder needs to completely design, build, and furnish a luxury home. Using the one-stop shopping approach means homeowners are assured a coordinated, smooth flow of work. Beause most people don't have the time, energy, or the expertise to chase all over town for the services needed, Dave Konkol Homes puts the entire process under one roof. People can come to the corporate office for the initial meetings, and rest assured that the Dave Konkol Homes team will take care of the whole package.

Dave Konkol Homes' team includes an architect, interior architect, interior designers, landscape architect, lighting consultants, experienced construction managers, an in-house real estate broker, and even custom mortgage broker services. There are so many components involved in custom designing and building a luxury home, that Dave believes this coordinated package is the best way to go. He has seen the chaos that can result if things fall through the cracks, and that's why his team provides the smooth flow of a one-stop, one umbrella approach.

From the beginning, Dave Konkol established what would make this company unique. The building blocks would be integrity and relationships. He trains his team to listen. The entire process flows from listening to what people say they want, and guiding them into a beautiful, successful completed home.

Dave Konkol Homes recognizes that although building beautiful homes is what they do every day, it's important to remember that for each one of their homeowners it's the only home that matters.

Visit www.DaveKonkolHomes.com for more information.

ABOUT THE AUTHOR

David A. Konkol, President

Dave Konkol Homes, Inc.

Dave Konkol grew up in a family of homebuilders and started his own roofing company at age seventeen. The roofing company was profitable enough to put Dave though college, and in 1984 he earned a bachelor's degree in Construction Management from the University of Wisconsin-Stout.

Upon graduation, Dave moved to Orlando, Florida, and has been building homes ever since. After working for a production builder for a few years, he started his own construction company in 1992 and began to define his niche by creating beautiful, custom-designed luxury homes starting at one million dollars.

Dave especially enjoys his family life. He is a devoted husband to Jo, his wife of fifteen years, and delights in being Dad to nine children, ages 3 to 10. He is an avid sports enthusiast and marathon runner. He also enjoys reading.

Serving God is another one of Dave's passions. He is an active leader in his local church and leads weekly men's group studies.

This is his first book.

SPEAKING OPPORTUNITIES

If you would like to invite Dave Konkol to speak to your group, please contact Dave at the address below. Please include dates, times, and location in your request.

Dave Konkol Homes, Inc.
1000 N. Maitland Avenue
Maitland, FL 32751
407-539-2938 (Office)

Dave.Konkol@DaveKonkolHomes.com

NOTES

NOTES

NOTES

NOTES

NOTES

NOTES

NOTES